Medical Device Quality Management Systems

MW00846361

Medical Device Quality Management Systems

Strategy and Techniques for Improving Efficiency and Effectiveness

Susanne Manz, MBA, MBB, RAC, CQA

BS Biomedical Engineering, Manz Consulting LLC,
Havre de Grace, MD, United States

ACADEMIC PRESS

An imprint of Elsevier

Academic Press is an imprint of Elsevier
125 London Wall, London EC2Y 5AS, United Kingdom
525 B Street, Suite 1650, San Diego, CA 92101, United States
50 Hampshire Street, 5th Floor, Cambridge, MA 02139, United States
The Boulevard, Langford Lane, Kidlington, Oxford OX5 1GB, United Kingdom

British Library Cataloguing-in-Publication Data
A catalogue record for this book is available from the British Library

Library of Congress Cataloging-in-Publication Data
A catalog record for this book is available from the Library of Congress

ISBN: 978-0-12-814221-9

For Information on all Academic Press publications
visit our website at https://www.elsevier.com/books-and-journals

Working together
to grow libraries in
developing countries

www.elsevier.com • www.bookaid.org

Publisher: Mara Conner
Acquisition Editor: Fiona Geraghty
Editorial Project Manager: Lindsay Lawrence
Production Project Manager: Kamesh Ramajogi
Cover Designer: Mark Rogers

Typeset by MPS Limited, Chennai, India

Contents

Acknowledgments

Thanks to all my colleagues over the years who helped me explore and develop my concepts of an effective and efficient quality management system.

Thanks to my family and friends who supported me and encouraged me throughout my career and while writing this book.

Introduction

In 1994, I changed jobs leaving the highly regulated aerospace industry at GE Aircraft Engines for another set of regulations. I was introduced to the current good manufacturing practices (cGMPs), which eventually became the Quality System Regulation for Medical Devices. I started a new role as a manufacturing facilitator at Ethicon Endo-Surgery, a subsidiary of Johnson & Johnson. With a degree in biomedical engineering, this was a perfect fit for me. I was so excited to work in the medical device industry. I had a passion for understanding the products and technologies, and how they worked with the human body. Not only did it allow me to use my education in biomedical engineering, but I felt like I was working on products that helped people. I was motivated by medical devices so important in relieving pain, diagnosing and curing diseases, and sustaining life. And I was so proud to be working in an industry that helped people and their families. I was deeply committed to creating products to improve the lives of customers and their families.

As my career at J&J progressed, I moved through various roles in manufacturing, engineering, quality, new product development, product support and postmarket surveillance, and compliance. I developed my skills, taking classes and other training, that enhanced my knowledge of the product life cycle for medical devices. I studied the concepts of quality and compliance for medical devices. I learned the difference between quality control and quality assurance. I learned about problem solving and process control. I was certified as a black belt and then a master black belt in Six Sigma. I became an executive consultant, helping executives develop their strategies, plans, and tactics for achieving quality and compliance goals.

As my career evolved, I relocated many times working at other J&J subsidiaries and later at Medtronic. With these changes, I learned about additional product types and technologies. I had various roles with worldwide responsibilities. These roles required me to travel extensively around the globe. And this continued to develop my awareness of various regulations in other countries. I also learned about the differing expectations and needs of customers, doctors, and patients around the world. Eventually, I left corporate life and became an independent consultant, helping other companies learn about quality and regulatory compliance.

These roles and experiences helped me to really understand all the details and nuances of the regulations. I deepened my understanding of customer needs and expectations. I once thought that compliance and regulations were a very boring topic. Now I see them as the means for ensuring that products are safe and effective.

It was not all rosy and I learned about the various things that can go wrong and how to deal with them. I have experienced many, many notified body audits and FDA inspections. I have seen times when things went very well and other times when things went very badly. I have seen a 38-page Form 483 (FDA's notice of observations). And I have seen small 483s that progressed to a warning letter. And I have seen how failure to effectively address issues progressed to a recidivist warning letter and eventually to a consent decree. These types of issues have enormous impact on a company's business success.

I made mistakes myself and I paid the price for those mistakes. I think back at all the things I wish I had known or all the things that I could have done better. But, I learned from those mistakes. Those mistakes improved my awareness of pitfalls. Those mistakes gave me greater insight into optimizing effectiveness and efficiency of the quality management system (QMS). I share my learnings with you in the hope that you can benefit from my mistakes.

I have experienced various quality problems and participated in the tough decisions of whether or not to do a recall. I have had to work closely with customers that were dissatisfied about product performance. I have personally felt the wrath of customers when products did not perform as intended. And even more importantly, I have had to deal with more serious quality issues. I have investigated serious injuries and deaths due to medical devices. Sometimes these quality issues progressed to legal disputes and I have had to give depositions about them at a later point in time. And I have seen the impact of those problems on customer satisfaction, market share, and good will. And when that happens, the business suffers. Employee moral suffers.

I also realized that most companies struggle to understand the requirements of the regulations. They have difficulty interpreting the intent of the regulations. And, therefore, they have great difficulty translating those regulations into an effective and efficient QMS. The mainstream quality practices, tools, and expectations are not always adopted by companies into their QMSs specific to medical devices. Many medical device companies simply do not have the knowledge, skills, and capabilities to create an effective yet efficient QMS.

I have also found that medical device companies often struggle to find the correct balance between effectiveness and efficiency. Lack of effectiveness results in compliance and quality problems such as 483 observations, warning letters, consent decrees, medical device reports, and recalls. We will explore this concept in Part I. And lack of efficiency leads to delayed product launches, poor utilization of resources, and lack of competitiveness and can also cause compliance and quality problems. We will explore this concept in Part II.

Efficiency is doing things right, effectiveness is doing the right things.

Peter Drucker

So, today I am an independent consultant with a mission to help other companies improve their quality and compliance. I help companies create and improve their quality systems and compliance results. I help companies who have had quality or compliance problems. I help with 483 and warning letter responses. And most of all I want to help medical device companies provide medical devices that are safe and effective for their customers. With this mission in mind, I would like to share my experiences and lessons with you.

My goals for *Lessons on Quality and Compliance*:

- Merge, consolidate, and organize concepts from various sources;
- Move from the "what" requirements of the medical device regulations and provide "how to" approaches for creating a QMS (Note: Product realization processes are not in the scope of this book and will be covered in a later companion volume);
- Share experiences and lessons;
- Provide examples and case studies;
- Provide structure for readers to use these lessons to shape their vision, strategy, and quality planning.

NOTE

Many of my lessons and examples are FDA centric but the concepts are applicable to other regulations.

Part I will cover the history of medical device regulation and how that shaped the regulations we know today. Prior to regulation, products were mostly ineffective and brought false hope and delays in proper treatment. And at the other extreme, unregulated products were often addictive, harmful, or even deadly. Con-men, quacks, and snake-oil salesmen used deceptive and misleading advertising techniques to deceive the public. Over time, regulations were put in place to protect the public. And the regulations continue to change even today. Expectations are ever-increasing as new products and technologies continue to change the medical device industry. We will explore the details of current expectations to "establish and maintain" a suitable and effective QMS that is compliant and results in improved product quality, providing safe and effective products. "Establish and maintain" requires a process approach and a defined QMS structure.

Part II will define characteristics of an efficient QMS as well as the relationship between efficiency and effectiveness. Of course, regulators and customers are concerned about quality system effectiveness. But, employees and shareholders have a stake in quality system efficiency as well.

Part III will identify and detail key roles and responsibilities for creating an efficient and effective QMS. Quality is not an organization! Every individual in the organization is responsible for quality. Key roles include management with executive responsibility, the management representative, process owners, and individual accountability. These are entwined in the fabric of a culture of quality. Part III will also define key capabilities or MEDICS essential for success.

Part IV will describe the role of the quality organization and how to bring value to the organization to earn a seat at the table. Concepts of quality and compliance have developed significantly from early days of quality control and quality assurance. World-class quality and compliance requires that medical device companies have a system of intelligence to define, measure, and articulate quality and compliance risks and opportunities. It is more common now for quality, compliance, and regulatory leaders to have a seat at the table to bring this intelligence to the table. Business success depends on analyzing, interpreting, and identifying global risks and leading risk reduction activities. Quality and compliance leaders facilitate interpretation and translation of regulations and constantly changing expectations, identify and mitigate risk, and use quality as a key competitive differentiator.

Part V will identify steps to translate vision and quality objectives into improvement strategies and more detailed plans. A quality system is a network of processes, people, and infrastructure that must work together in a cohesive, aligned manner to realize quality objectives. An efficient and effective QMS takes ongoing effort to look for signs of misalignment and take appropriate action to drive stakeholder alignment.

Part VI will provide an overview of critical improvement methodologies and tools, including corrective and preventive action, root cause analysis, as well as inspection preparedness.

This book is for quality, compliance, and regulatory professionals and leaders who want to find their place at the table to influence business strategy and success. It is for professionals who want to understand the current quality situation and maturity level to create a vision and strategy for improvement.

This book is for process owners and functional leaders who want to improve quality and compliance to drive more efficient processes and predictable results. This book is for process owners who want to be more aware of their roles and responsibilities for quality and compliance results and how to meet regulatory expectations. A process that does not consistently meet regulatory requirements is not effective. And an ineffective process will never yield predictable business results! Improving process effectiveness can reduce risks and eliminate nasty surprises and delays.

And this book is for management with executive responsibility who want to better understand their role and responsibilities to establish and maintain a suitable and effective QMS. This book is for management that wants deeper understanding of quality and compliance risks and how to identify and react to them. This book is for management that wants quality and compliance to enable business success.

This book is for anyone in the organization that wants to learn about overall quality and compliance. It is for anyone that wants to shape and improve processes to make sure that the products you work on help your customers, reduce their pain, get them back on their feet and home to their families.

This book deals with quality management topics in a summary manner and provides a road map for improved efficiency and effectiveness of the QMS. It is impossible to deal with these topics in exhaustive detail in one single book. Readers are encouraged to explore topics in greater detail. Use the reference documents in the bibliography for further detail. Be sure to reference the actual regulations to confirm your actions and decisions regarding your specific QMS.

An effective
quality system

Regulatory requirements

Today, medical devices are highly regulated with a long history of regulations going back to the *Pure Food and Drug Act* of 1906. Prior to this, medical products were unregulated, promised to cure a wide variety of vague ailments, and were often ineffective or even hazardous. Patent medicines originated in England and were associated with proprietary "medicines" or concoctions that arrived in North America with the first settlers. The term "patent medicine" is a bit confusing and does not always mean medicines that have been patented. The phrase "patent medicine" comes from medical elixirs of the late 17th century when those who found favor with royalty were issued letters patent that allowed the use of royal endorsement for advertising.

Patent medicines, also known as nostrums, were products that were proprietary in nature, unproven, and marketed directly to the public. Patent medicines were sold with interesting names, bogus claims, and using advertising games. Pioneer (aggressive) advertising techniques with distinctive trademarks and packaging were common. Products were promoted as cure-alls and panaceas for all sorts of ailments. Entrepreneurs began to bottle and sell "old family recipes" with no proof of efficacy or safety. Networks of unscrupulous traveling salesmen sprang up and turned patent medicines into big business. Profit-making took priority over safe and effective products.

These quack medicines were often not effective and did not contain the ingredients specified. Even worse, they were sometimes deadly. Often, they had high (and undisclosed) alcohol content which may explain why they were so popular. Ingredients were unregulated and could include those that were addictive and lethal. They often contained addictive ingredients such as morphine, opium, cocaine, or heroin. Mercury and arsenic were found in some products. One product containing opium, alcohol, and chloroform together promised to cure everything from toothache in 5 minutes to colds in 24 hours and deafness in 2 days! Even worse, patent medicines were frequently marketed as "infant soothers" and "teething cordials" for fussy, teething, or colicky babies, but caused tragic results. Mrs. Winslow's Soothing Syrup claimed that it was "likely to soothe any human or animal."

Newspapers became filled with advertisements promising quick, cheap cures for all ailments both dreadful and mundane. Advertisements for patent medicines were everywhere and referred to vague disorders such as female complaints and

Medical Device Quality Management Systems. DOI: https://doi.org/10.1016/B978-0-12-814221-9.00001-4

dyspepsia. Advertising cards (trade cards) were colorful, imaginative, and displayed distressing before pictures (when the card was closed) and happier after pictures when the card was opened. Testimonials were a common advertising technique.

Sellers made broad claims about patent medicines curing a variety of ailments. These concoctions were sold to the unsuspecting public with no controls. Salesmen often used innovative techniques to attract attention to their sales offerings. At the 1893 World Expo in Chicago, Clark Stanley, the "rattlesnake king," caused a stir by publicly killing rattle snakes and processing their bodies to collect snake oil to "stop pain, reduce swelling, remove inflammation, and oil dry joints."

NOTE

A visit to the Smithsonian Institution's National Museum of American History in Washington, DC or the Hagley Museum in Wilmington, Delaware provides many fascinating examples of patent medicines and their colorful bottles, advertisements, and quack claims. Items include:

- Stetketee's Neuralgia Drops
- Aubergier's Pastilles of Lacucarium
- Kickapoo Indian Sagwa Renovator
- Renne's Magic Oil
- One Minute Cough Cure

By the early 1900s, some physicians and medical societies were recognizing problems and becoming critical of these medicines. At this time, there were estimates of 50,000 patent medicines being manufactured and sold in the United States. Journalists began exposing injuries as well as proprietary formulas. American journalist and author, Samuel Hopkins Adams wrote a series of 11 articles in 1905 to 1906, *The Great American Fraud*, that called attention to these issues.

Other changes were occurring, and Upton Sinclair described the unsanitary conditions of meat-packing plants in *The Jungle*. This led to widespread public outrage and demand for regulation of food and drug products. With the strong support of President Theodore Roosevelt, the *Pure Food and Drug Act* was passed in 1906 and started an era of regulation to ensure that medical products were both safe and effective.

In 1917, the Bureau of Chemistry (later known as the Food and Drug Administration or FDA) seized a shipment of Stanley's Snake Oil and found that it contained no snake oil at all. It was mainly mineral oil, red pepper, and turpentine. Clark Stanley was fined $20. The term "snake oil" was now well-established as a quack concoction marketed as a patent medicine. Even today, the term "snake-oil salesman" still conjures the image of a seedy salesman peddling his wares on a soapbox in a traveling show.

In 1911, the American Medical Association published *Nostrums and Quackery*, describing hundreds of products linked to serious injuries and deaths. Mrs. Winslow's Soothing Syrup was featured in a section titled "Baby Killers." By this time, there were also concerns about misleading advertising, quackery, and mechanical products (the early medical devices). An entire chapter was devoted to the advertising techniques of con men and quacks.

A chapter titled *Mechanical Fakes*, voiced concerns about unproven new medical devices. "It is sometimes hard to decide which is greater—the impudence of the quack or the credulity of his victims. The comparative ease with which the medical faker is able, by the most preposterous of claims, to separate the trusting from their money indicates the enormous potentialities in advertising. It might well be supposed that an individual who set out to sell, as a panacea for all the ills of the flesh, a piece of brass pipe with one or two wires attached to it, would commercially speaking have a hard and rocky road ahead of them. But such supposition would be incorrect. Not only would the enterprising faker find customers for his gas pipe, but there would be such a demand for this most inane of 'therapeutic' device that two or three imitators would immediately enter the market." This described the Electropoise device, originated by "one Hercules Sanche, who modestly described himself as the 'Discoverer of the Laws of Spontaneous Cure of Disease'." The book goes on to state that, in order to minimize legal issues and maximize financial gain, "Sanche finally hit on a device that was negatively harmless—and positively worthless—and yet theatrical enough to make the purchaser feel that he was getting something for his money." Advertisements for Electropoise made wild claims, even saying that it could cure previously determined incurable diseases. It was time for improved regulation to protect the public health.

The *Pure Food and Drug Act* of 1906 regulated only drugs and food and did not include medical devices. It was replaced in 1938 by *The Food, Drug, and Cosmetic Act (FD&C Act)*. The *FD&C Act* (sometimes referred to in medical device regulations as *the Act*) provides the framework for the regulations we know today. The *FD&C Act*, along with various amendments, forms our current regulatory requirements. The *FD&C Act* included the following provisions:

- Extended control to cosmetics and medical devices
- Gave the FDA formal authority to inspect factories, warehouses, or establishments
- Required drugs to be shown as safe before marketing (premarketing conditions)
- Added injunctions to the previous penalties of seizures and prosecutions
- Clarified the concept of prescription vs over-the-counter drugs in the *Durham Humphrey Amendment* of 1951
- Required, via the *Kefauver Harris Amendment* of 1962, efficacy as well as safety data before a drug could be marketed and instituted stricter agency controls. It also transferred regulation of advertising from the Federal Trade Commission to the FDA. And it established *Good Manufacturing Practices (GMPs)*.

At the time the FD&C Act was approved, medical devices were very simple instruments such as scalpels and stethoscopes. By the 1960s and 1970s, there was a significant increase in the number, complexity, and risk of medical devices. But, the FDA's limited resources were focused on dealing with quack devices. The Cooper Commission report of 1970 determined that more than 700 deaths and 10,000 injuries were attributed to medical devices. This resulted in the *Medical Device Amendments* to the *FD&C Act* signed into law by President Ford in 1976. This Act included:

- Authority for FDA to regulate and control the release of medical devices into interstate commerce
- Requirements for device listing, establishment registration, and adherence to GMPs
- A three-tiered classification scheme for products based on risk
- Provided procedures for authorization for distribution which include 501(k) premarket notification and premarket approval (PMA);
- Provided authority for the FDA to ban devices

The *Safe Medical Devices Act of 1990 (SMDA)* included:

- The first requirements for medical device reporting (section 803) to the FDA
- Requirements for device tracking and post-market surveillance
- Allowed the FDA to temporarily suspend approval of PMAs for new products
- Allowed civil money penalties for violations
- Added Humanitarian use provisions
- Provided FDA with the authority to add preproduction design controls to the cGMP (current Good Manufacturing Practices) Regulation.

The SMDA defined a medical device as "any instrument, apparatus, implement, machine, contrivance, implant, in vitro reagent, or other similar or related article, including a component part or accessory which is:

- Recognized in the official National Formulary, or the United States Pharmacopeia, or any supplement to them,
- Intended for use in the diagnosis of disease or other conditions, or the cure, mitigation, treatment, or prevention of disease, in man or other animals, or
- Intended to affect the structure or function of the body of man or other animals,
- And which does not achieve its primary intended purposes through chemical action within or on the body of humans or other animals and which is not dependent on being metabolized for the achievement of any of its primary intended purposes."

This definition provided a clear distinction between medical devices and other regulated medical products such as drugs and biologics.

The *Medical Device Amendments of 1992* clarified medical device reporting requirements with expectations for user facilities, manufacturers, importers, and distributors.

In 1996, the cGMP (current Good Manufacturing Practices) was revised to add design control requirements and achieve some alignment with international standards. This resulted in the *Quality System Regulation (QSR)* found today in the Code of Federal Regulations (CFR) Title 21 Part 820. The preamble to the QSR provides very useful information and interpretation of the FDA's rationale and thinking at the time. It is recommended that all quality and compliance professionals read the entire preamble to enhance their understanding of the QSR.

This short history gives a framework for our current medical device regulations in the United States. It helps understand the issues that shaped the laws we know today. Many details in the laws were in response to specific events or problems that highlighted the need for regulation.

As a consumer, I am personally grateful for the all work and effort it has taken to create the laws and the regulations aimed at protecting the public health. I know that these regulations ensure the safety and efficacy of medical devices. And as a quality and compliance professional in the medical device industry, I find the history fascinating in understanding the complexities of that industry. The history provides context and interpretation of the specific issues that led to the laws and the changes over time. The regulations continue to develop and throughout this book we will discuss current issues and future trends.

Many medical device manufacturers also design, manufacture, sell, distribute, and service their devices in other countries. It is impossible to exhaustively cover the laws and regulations for other countries. But, *International Standard ISO 13485 Medical devices—Quality management systems—Requirements for regulatory purposes* is an important standard to understand. ISO 13485 is the basis for complying with the *Medical Device Directive* (MDD) enacted by European Parliament. Compliance with the MDD is signified with the CE (European Conformity) mark on products. CE marks are granted by notified bodies (e.g., TüV, BSI, or KEMA), private companies accredited by country governmental bodies. The 2003 version of ISO13485 is still in use at the writing of this book. However, many medical device manufacturers are in the process of transitioning to the 2016 version to meet the required 2019 deadline. We will discuss changing expectations for the medical device industry further in Chapter 2, Increasing Expectations.

There are many similarities and a few differences between the QSR and ISO 13485. We discusses some important distinctions throughout this book. The *International Medical Device Regulators Forum (IMDRF)* is an important organization that seeks to "accelerate international medical device regulatory harmonization and convergence." IMDRF was started in 2011 and now includes Australia, Brazil, Canada, China, Europe, Japan, Russia, Singapore, the United States, and the World Health Organization. IMDRF built on the work of a previous organization called the *Global Harmonization Task Force (GHTF)* on Medical Devices. The IMDRF website is a very useful resource. It contains many important and useful guidance documents, including GHTF documents.

> **TIP**
>
> It is recommended that medical device manufacturers monitor the activities of IMDRF (and other applicable regulators) on an on-going basis to understand regulatory issues, priorities, and harmonization activities.

Medical devices are classified to determine regulatory requirements. In 1976, the FDA took all known devices and organized them into 19 classification panels. In 1990, the number changed to 16. These panels are grouped according to medical specialties such as cardiovascular, neurology, and orthopedic devices. Within these panels are approximately 1,700 types of devices. In Canada, Europe, and other countries, there are similar but somewhat different classification schemes. Product classification is important as it determines:

- Risk level
- Activities that need to occur before you market your product (premarket requirements)
- The required level of controls within your quality management system (QMS)

Class I devices are those for which general controls are considered sufficient. General controls include:

- Prohibition against adulterated or misbranded devices
- Premarket notification [510(k)] requirements
- Banned devices
- GMPs
- Registration of manufacturing facilities
- Listing of device types
- Record keeping
- Repair, replacement, and refund

Class II devices are higher risk product that also require special controls such as design controls, mandatory performance standards, and other product-specific controls.

Class III devices also require premarket approval to ensure their safety and effectiveness. Premarket approval is the process of scientific review to ensure the safety and effectiveness data of new products. These products include devices that are life sustaining or life supporting. Additional controls such as traceability by serial number are required.

To comply with the above regulations, a medical device manufacturer must establish a Quality Management System or QMS. The FDA defines a quality system as the organizational structure, responsibilities, procedures, processes, and resources for implementing quality management. "Each manufacturer shall establish and maintain a quality system that is appropriate for the specific medical device(s) designed or manufactured, and that meets the requirements of this part."

The purpose of a QMS is to:

- Ensure devices are safe and effective
- Satisfy customer quality requirements and expectations
- Ensure labeling and product information is complete and correct
- Comply with regulatory requirements

First and foremost, a QMS must be *effective*. It must achieve the intended purpose. An *effective* QMS must ensure compliance with applicable regulations. And even more importantly, it must ensure the quality of the medical devices so that they are safe and effective. Quality of the medical devices includes the device itself plus packaging, labeling, and instructions. For the remainder of this chapter, we will explore the concepts of an effective QMS.

The regulation says that the QMS must be "appropriate" for the specific medical device(s) manufactured. This means that you need to understand the classification and risk level of your medical devices along with applicable requirements and standards. Special controls and standards can be found by searching the device classification panels for definitions and requirements. The FDA also has a database of international consensus standards that it recognizes (e.g., *ISO 14971 Medical Devices—Application of Risk Management to Medical Devices*, 2007). You will also need to consider your company organizational structure, complexity, medical specialties, sales markets, etc. to determine what is specifically applicable to and "appropriate" for your QMS. Regulations are always considered appropriate unless you can clearly demonstrate why not (e.g., a procedure for installation is not applicable if you do not design or manufacture products requiring installation).

From an FDA regulatory standpoint, two important considerations for quality are (1) the prohibitions against misbranding and (2) adulteration of medical devices. These long-standing concepts stem from the original *Pure Food and Drug Act* of 1906.

Adulterated devices include:

- The product or container is composed, in whole or in part, of any poisonous or deleterious substance
- Contains unsafe colorings or additives
- Differs in strength or purity from that claimed
- Does not comply with performance standards
- Fails to conform to requirements for premarket approval
- Is a banned device
- Fails to comply with investigational device exemptions
- Is in violation of GMP requirements

Violations of the GMP requirements (the last item) are common in inspections and result in product that is considered adulterated. One of the most frequent comments seen in warning letters is "your firm's devices are adulterated within the meaning of the Act in that the methods used in, or the facilities or controls, used for, their manufacture, packing, storage, or installation are not in conformity with the current good manufacturing requirements of the QSR found at Title 21, Code of Federal Regulations (CFR), Part 820."

Many warning letters include citations for misbranding of products. Misbranded devices include:

- The label is false or misleading;
- The packaging does not bear a label containing the name and place of business of manufacturer;
- The label fails to contain an accurate statement of quantity.
- Any word or statement not prominently placed or clearly stated so as to be read and understood by ordinary individuals under customary conditions of purchase and use;
- It is for use in man and contains habit forming substances and does not provide a warning;
- The label does not bear adequate instructions for use;
- It is dangerous when used as labeled;
- It contains false or misleading advertising;
- The device is distributed without appropriate 510(k) submission.

From these regulations, it is clear how expectations were developed from the original *Pure Food and Drug Act* of 1906. For example, the wording reflects the specific issue of Patent medicines containing habit forming additives such as morphine or opium. Regulations against false or misleading advertising are reflective of gross issues like the fake medical device, Electropoise.

Without understanding the history of medical device regulation, one would not easily understand the expectation that medical devices do not contain habit forming additives. Although some of the wording seems a bit old-fashioned, it is perfectly understandable within the context of early, unregulated products.

ENFORCEMENT

The Food and Drug Administration or FDA (also referred to as the agency) is a federal agency of the United States Department of Health and Human Services. The FDA is responsible for protecting and promoting public health through the control and supervision of food, tobacco, drugs, radiation emitting products, medical devices, and other products. The FDA is empowered by Congress to enforce the Food, Drug, and Cosmetic Act. The FDA has many mechanisms at its disposal to enforce the regulations. These include:

- Notice of Inspectional Observations (Form 483) which are issued by an FDA investigator. This is the form used to report all significant objectionable conditions noted during an inspection. These are legally considered as allegations of violations but do not constitute a conclusion or final agency action. Some, but not all, 483 observations will escalate to a warning letter or more serious enforcement action. An inadequate response to a Form 483 may also escalate to a warning letter. See Chapter 16, FDA Inspection Readiness, for more information.

- An untitled letter from the FDA may raise concerns or can also be used to gather additional information.
- A Warning Letter is written communication designated as a "warning letter" notifying a firm that the agency considers one or more products, processes, of other activities to be in violation. It may precede more serious enforcement action. A warning letter is considered a public record and is posted on the FDA website for public viewing.
- A recidivist Warning Letter is a type of warning letter that lists additional concerns when a firm has a pattern of repeat issues.
- Regulatory meetings ensure communication and clarity.
- A Consent Decree of Permanent Injunction is a negotiated settlement between a firm and the FDA that is filed with and approved by a court.
- Withdrawal of marketing approval is a technique to revoke approval for new 510(k) or PMA products. This can have serious negative consequences for companies who then cannot launch new products. Competitors take advantage to seize market share.
- Mandatory recall may be required per 21 CFR 810, Medical Device Recall Authority, which describes the procedures the FDA will follow in exercising its authority. Such an action may be undertaken when a firm refuses to issue a voluntary correction or removal. Any responsible company will avoid this by voluntarily conducting a recall.
- Detention or Refusal of Imported Products is an administrative act, where the FDA requires that products are held or refused entry altogether. It may be preceded by an import alert or may be a result of import inspections.
- Seizure allows the FDA to take possession of product already released to interstate commerce because it is in violation of the law. FDA initiates seizure by filling a complaint with the US District Court. If approved, the judge issues a seizure order, at which point, Federal Marshalls take possession of the goods. Seizure may be specific to a lot, be open-ended, or a mass seizure of all products at an establishment.
- An injunction is a civil action authorized by a court against an individual to stop or prevent violation of the law, such as halt the flow of violative products into interstate commerce.
- Criminal prosecution can be taken against a company or individual, charging violation of the law.
- Civil money penalty is a noncriminal action assessed by the FDA or courts for violations of the *FD&C Act*.

As we can see, there are increasing levels of actions that the FDA can take to enforce compliance of the law. Establishing an effective QMS is a difficult task, and it is common for an FDA inspection to result in a Form 483 with observations. But, a medical device company must respond promptly and effectively to avoid escalation of enforcement penalties. Product quality problems deserve the highest level of attention and management commitment to address them. If the company does not respond and address issues in a reasonable manner, the FDA can and will escalate to a warning letter, consent decree, and other more punative steps.

We often hear news about warning letters and some consent decrees. But, very seldom do we hear about civil penalties and criminal prosecution. Some notable instances are discussed in Chapter 7, Quality is not an Organization, regarding management responsibility.

An ineffective QMS can result in serious quality or compliance issues such as recalls, 483 observations, warning letters, and law suits. Serious issues can become publicly known and result in customer dissatisfaction, declining sales, reputation damage, and poor business results. The impact of these problems can be very significant. Responding to a 483 and taking appropriate corrective actions can take significant amounts of time and resources. Some companies spend hundreds of thousands or even millions of dollars dealing with compliance issues. Small companies often need expert (and expensive) consultants to provide an adequate response and corrective actions. A warning letter or a consent decree can be even more significant. I have known companies that had to halt all development activities and new product launches to respond to a warning letter. I've known companies that have diverted all resources from new product development to dealing with issues such as remediation of old design history files, complaint management, and medical device reporting corrective actions. Inability to launch new products can be a crippling blow to any company. Sometimes companies are required to halt production or close plants. In the case of a recidivist warning letter, the FDA can require a medical device manufacturer to bring in (at their own expense) an approved external consultant to do annual inspections and provide the certified results to the FDA. I have known prominent CEOs that have had to go to the FDA headquarters in Silver Spring, Maryland on a quarterly basis to personally report on their progress.

A recall can be a tremendous blow to a company and can cost tens of millions of dollars. It can have a devastating effect on company sales, reputation, brand image, and stock price. But, even more significant than these compliance issues is knowing that medical products can cause harm, serious injury, or even death for customers. I think most medical device companies and their employees do not intend for these things to happen. But lack of awareness of the regulations can result in terrible consequences for customers, patients and their loved ones, and medical professionals.

> **WARNING**
>
> Always be aware of signs that your employees don't understand or respect the regulations. As noted in the regulatory history for medical devices, the regulations exist because of specific issues and problems. Personnel who consider the regulations burdensome may need additional training and awareness of the history and reasons for the regulations. They need training to understand the relevance of their role and importance on product quality.

Of course, a recall or a warning letter only provides lagging information about problems with your QMS. Medical device manufacturers are required to do certain activities proactively to make sure that their QMS is suitable and effective. These include requirements for conducting internal audit and management review. Internal Audit and Management Review will be discussed in Chapter 8, Capabilities and MEDICS for an Effective QMS.

It is important to be alert for some of the warning signs that your QMS is ineffective:

- There are serious, numerous minor, and/or repetitive failures to comply with regulations and internal procedures.
- Serious, numerous minor, and/or repetitive nonconformities are found in internal audits.
- Any nonconformities are found during external inspections.
- Serious and/or repetitive quality issues such as MDRs (medical device reports) and recalls.
- Metrics and dashboards show negative trends or failure to meet goals and address problems.
- Resources are spent more on fixing issues rather than preventing issues. For example, I once worked with an organization that had a department called "Recall Management" devoted only to dealing with the company's many recalls.
- Unacceptable personnel behaviors aimed at avoiding, hiding, minimizing, diluting, or neglecting quality and compliance issues are observed.
- *Corrective and Preventive Actions (CAPAs)* are routinely late, incomplete, or ineffective.

We will explore metrics and other key capabilities (MEDICS) for monitoring the health of your QMS in Chapter 8, Capabilities and MEDICS for an Effective QMS.

> **WARNING**
>
> If you've had an FDA inspection (or notified body inspection), do not be fooled into thinking that an absence of a Form 483 (or major nonconformity) means that your QMS is acceptable. An FDA inspection is only a snapshot at a given point in time. It is possible that the FDA may not find any
>
> *(Continued)*

> **WARNING (CONTINUED)**
>
> nonconformities in a given individual inspection. I have seen companies have a clean inspection one time, followed 2 years later by nonconformities serious enough to warrant a warning letter! You must have your own internal mechanisms to thoroughly and proactively evaluate the health of your QMS (see Chapter 8, Capabilities and MEDICS for an Effective QMS, on MEDICS).

Allowing, managing, and responding to an FDA Inspection is a regular part of doing business in the medical device industry. Consumer Safety Officers, more commonly called investigators, inspect medical device manufacturers. Key success factors for external inspections include:

- Knowing your current compliance status with FDA and other regulatory bodies
- Understanding your risk areas including both quality and compliance
- Awareness of common or systemic risks throughout your organization
- Understanding your inspection history, results, and corrective action status
- And having a plan and process to manage the inspection (see Chapter 16: FDA Inspection Readiness).

If you have developed an effective and efficient QMS, you should always be "inspection ready." There should be no need for panic and a flurry of last-minute activities. An essential tool to be inspection ready is understanding the FDA's current inspectional process known as *Quality System Inspection Technique (QSIT)*. The FDAs *Guide to Inspections of Quality Systems* is available on the FDA website and is essential reading for Quality, Compliance, and Regulatory personnel at medical device manufacturing companies. The QSIT guide identifies seven quality subsystems and related satellites. Of these, four are commonly inspected (Management Responsibility; CAPA; Design Controls; and Production & Process Controls) (Fig. 1.1).

Throughout this book, we will refer to the QSIT manual for development of your QMS. The QSIT manual provides valuable information for developing your QMS. See Chapter 16, FDA Inspection Readiness, for more information on inspection preparedness.

Conclusion: The current regulations for medical device companies were shaped by a long history of specific instances and issues going back to unregulated patent medicines. Understanding this history is beneficial for medical device companies to develop an appropriate and effective QMS. The Food, Drug, and Cosmetic Act, with amendments, is the foundation of our current regulations and is enforced by the FDA. Failure to create an effective QMS can result in compliance and/or product quality violations. The FDA has many mechanisms at its disposal to enforce regulations.

This short history can guide and inspire efforts today to develop effective and efficient QMSs to provide safe and effective products for customers using scientifically based evidence.

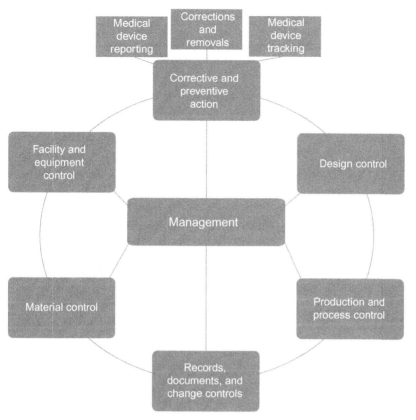

FIGURE 1.1

Quality system inspection technique subsystems.

FUTURE TRENDS

Regulations will continue to evolve due to new issues that arise from the increasing complexity and technology in medical devices. Software is an example of the changing nature of medical devices. With a multitude of technology platforms, ease of access (internet, cloud, apps), and communication methods, software is increasingly used as a medical device. Additionally, many medical devices now have imbedded software and/or the ability to be programmed externally. There are several types of wirelessly reprogrammable, implantable devices on the market now. These may be life-sustaining devices. In an episode of the popular fictional TV show *Homeland*, a terrorist assassinated the vice-president of the United States by hacking his pacemaker. Although, there are details of this TV scenario that were unrealistic, it does remind us that medical device manufacturers must now think of both device security and privacy implications. In 2016,

IMDRF released a proposed document on *Software as a Medical Device (SaMD)* to establish common and converged understanding of clinical evaluation and principles for understanding and principles for demonstrating safety, effectiveness, and performance of software intended for medical purposes.

Today we have highly complex medical devices, sometimes in combination with drugs or biologics. We have very large, complicated, and durable equipment such as MRI or CT scanners. Medical devices are commonly implanted into the human body and are lifestyle enabling or life sustaining. Some medical devices are active, relying on a source of electrical energy or power. We have digital health technologies with rapid lifecycle iterations. And we can expect the complexity and importance of medical devices to continue to grow. Personalized medical devices, minimally invasive devices, advanced combination products, bioelectronics, novel materials, robotic surgery, nanotechnology, collaborative care, new manufacturing techniques, reuse and reprocessing of single-use devices, and 3D printing are all trends that may result in additional changes to medical device regulations. For example, the FDA already released "leapfrog" guidance on additive manufacturing (3D printing) to provide initial guidance on this emerging technology. This includes expectations for patient-matched design, effects of imaging on fit, cyber-security, and personally matched information.

TIP

It is important to monitor changes in medical device technology and regulatory expectations, and standards. Industry associations such as Advamed, AAMI, IMDRF, and MDIC are good sources of information. Ensure that you have a process for or person in charge of monitoring the external environment. Significant changes should be identified in your QMS strategy.

Increasing expectations

2

In a world of continuously increasing expectations and immediate visibility to issues, medical device manufacturers need to be aware of the importance of building a reputation of ethical behavior, compliance, quality, and value. They must earn the trust of their customers. One recall, one warning letter, or one news story on TV can destroy years of good results. Social license is granted by customers, patients, doctors, hospitals, and easily taken away. Corporate social responsibility is impacted not only by quality and compliance but also by environmental, sustainability, and social expectations.

Employees and shareholders also care about the environment and sustainability. They want to work in diverse organizations that respect employee needs and safety. Employees are motivated knowing that the products they work on help people. They are more satisfied working for socially responsible organizations with a reputation of quality, compliance, and value. Shareholders not only expect a return on investment, they want to know that it was the product of an ethical, compliant, and quality driven effort. A vigorous, effective quality management system is essential to creating that reputation.

Over the last few years, there have been ever-increasing expectations about quality and compliance from regulators around the world. One change is the concept that compliance with the regulations, on its own, is not enough to ensure safe and effective products that meet customer needs and expectations. There is a difference between compliance and quality. Most people define quality more broadly than compliance. Compliance is defined as meeting regulatory requirements. Quality is defined as products and service that meet customer needs and expectations. Although the regulations are intended to ensure that compliance results in quality outcomes, on a practical level this is not always the case. Science-based evidence is necessary to demonstrate the device is not only effective, but the most effective therapy.

In 2011, the Food and Drug Administration's (FDA's) Center for Devices and Radiological Health (CDRH) launched the *Case for Quality* by presenting the report *Understanding Barriers to Medical Device Quality*. This document was based on reports that the regulatory focus on compliance alone was not resulting in improved product quality and protection of the public health. Since then, the FDA has taken active steps to promote the "Case for Quality."

Regulators in Europe have also made changes to regulations. ISO 13485:2016 puts a much stronger emphasis on the customer than does the Quality System

Medical Device Quality Management Systems. DOI: https://doi.org/10.1016/B978-0-12-814221-9.00002-6

Regulation. Customer feedback includes the premise that you are proactively seeking feedback from customers rather than waiting for complaints to come to you. It is a much more proactive approach and provides leading indicators of customer satisfaction. Changes to ISO 13485 in 2016 placed a stronger emphasis on risk management and taking actions that are commensurate with risk.

CASE STUDY

In December 2011, authorities in France advised 30,000 French women to have potentially defective breast implants removed. The manufacturer, Poly Implant Protheses (PIP), had used industrial grade silicone in the implants instead of medical grade silicone. There was mass confusion in public media about the issue and how to deal with it. Women with implants, their doctors, and even other regulators were not able to get clear, scientifically based information, or advice about the extent and nature of the issue. More than 400,000 women in 65 countries were believed to have received the faulty implants. They struggled with the choice of having another operation to have the implants removed or leave them in place with a possible rupture. There were 4,000 ruptures reported. Over 125,000 women made the difficult choice to have the implants removed.

Jean-Claude Mas, the founder of PIP, was sentenced to 4 years in prison, fined 75,000 euros, and banned for life from working in medical devices. The PIP scandal highlighted the differences between the European regulations and the US regulations. It also highlighted the complexities of the European system and the many organizational bodies (countries) involved. It was clear that regulators were unable to efficiently or effectively share information with each other. Without the ability to get information from other regulators, they were unable to provide information and recommendations to the public they were dedicated to protecting.

The PIP breast implant scandal had far reaching effects. Since then, regulators have looked for opportunities to improve and to share information with each other. The European Commission and member states created an action plan based on the PIP breast implant scandal that focused on control of notified bodies; market surveillance; coordination of vigilance; communication; and transparency. They have worked to make changes to the regulations and expectations. The European Medical Device Regulation (MDR) was revised in 2017.

In addition, regulators have worked together to harmonize activities, share information and best practices, and leverage resources. Formed in 2011, *International Medical Device Regulators Forum* (IMDRF) is a regulator only forum to advance harmonization. It is a voluntary group of regulators around the world that have come together to "build on the strong foundational work of the Global Harmonization Task Force (GHTF) on Medical Devices, and to accelerate international medical device harmonization and convergence." Current members include Australia, Brazil, Canada, China, Europe, Japan, Russia, Singapore, South Korea, and the United States working with the World Health Organization (WHO).

Organizations like IMDRF facilitate sharing information with each other thus preventing some of the communication issues that were observed in the PIP breast implant scandal. In addition, they are able to harmonize methods and leverage each other to conduct regulatory inspections of manufacturers through the *Medical Device Single Audit Program* (MDSAP). In addition, MDSAP reduces the burden on industry of having multiple regulatory inspections each year and replacing them with a single MDSAP inspection.

Visibility of quality and compliance issues has changed significantly over the last few years. Increased visibility of product quality data makes it more important than ever to for medical device manufacturers to be able to differentiate their product quality and compliance results. New sources of information include:

- Social media such as Twitter and Facebook;
- Comparative product quality data;
- National and international registries such as MedSun for medical devices to study outcomes;
- National Evaluation System for Health Technology Coordinating Center (NESTcc) "seeks to support the sustainable generation and use of timely, reliable, and cost effective Real-World evidence throughout the medical device lifecycle";
- FDA is partnering with other federal agencies such as NIH (National Institutes of Health), CDC (Centers for Disease Control and Prevention), and CMS (Centers for Medicare and Medicaid Services) to monitor and analyze data.

Product Quality Metrics will be used increasingly in the future to add additional data that paints a clearer picture of product quality risk. Product Quality Metrics are already being piloted and shared with the FDA to provide information that adds transparency to pre- and postmarket data already gathered. These metrics may be used in the future to understand risk and prioritize regulatory inspections.

In 2013, the FDA issued its final rule on Unique Device Identifiers (UDIs). UDI requirements are being phased in over 7 years. The increasing use of UDIs facilitates the collection and analysis of data on quality of outcomes related to medical devices. It allows detailed analysis and segmentation of quality outcomes data by device type, manufacturer, and more. UDI, product metrics, and electronic submission of data will significantly streamline and enhance the ability of the FDA to collect and analyze the data in its efforts to protect the public health.

Electronic submission of data to the FDA is the norm now. The FDA required electronic reporting MDRs (*Medical Device Reports*) in 2015. Electronic data submissions reduce data entry burden, allow faster analysis, and facilitate analysis and comparison of data throughout the product life cycle (pre- and postmarket information). Regulators have increased their abilities to receive information electronically thus enhancing their speed and ability in analyzing vast amounts of data. Medical device manufacturers need to be doing the same enhancements to speed and ability. Medical device manufacturers need to use these new abilities to look forward and predict risks. They need to be able to understand their own product quality and react to faster than regulators.

On December 14, 2016, the FDA issued final guidance on "Emerging Signals." This guidance provides information on how the FDA will use information on emerging quality problems to notify the public. The goal of this guidance was to provide the FDA's current thinking on how to ensure that the public gets accurate, science-based information about medical products to maintain and improve public health. The FDA described the factors for consideration, the process, timelines, mechanism for public notification, etc. This Emerging Signals guidance reflects a significant improvement in transparency and communication to the public and other regulators since the PIP breast implant scandal.

The FDA continues to advance its efforts to protect the public health. FDA Commissioner, Scott Gottlieb, MD included the following items in his 2018 Strategic Policy Roadmap (reference the FDA website):

- Adopt a team-based approach to regulation of medical devices.
- Install a new Total Product Lifecycle office.
- Develop and enhance the use of *in silico* tools and models to evaluate device performance and patient outcomes as part of the Medical Device Innovation Consortium.
- Pursue opportunities for patients to get access to information that can better inform medical decisions.
- Foster innovation in digital health technology tools that can help better inform and empower consumers.

Fig. 2.1 summarizes the improvements of regulators to collect, analyze, and react to information necessary to protect the public health. Expectations continue to increase for medical device manufacturers. There is an increasing emphasis on

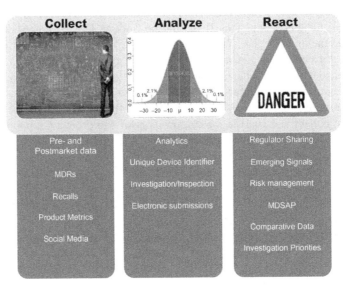

FIGURE 2.1

Increasing regulator capabilities.

quality outcomes and scientific evidence. Compliance is only the baseline. Customer focus and effective risk management are increasingly important.

Of course, sometimes regulators get things wrong. They face challenges to find the fine line between speed to market and ensuring complete safety of new medical devices. They are frequently criticized for adding excessive burden to requirements and slowing the delivery of innovative new products. But, when they try to speed products through, they are criticized for not being effective. VIOXX is a well-known example of the FDA receiving criticism for prematurely approving a product. Ultimately, Merck, the manufacturer of VIOXX, withdrew the product from the market due to elevated cardiovascular risk.

In conclusion, regulators have made strong efforts to increase their capabilities to gather, analyze, and react to information, fulfilling their mandate to protect the public health. These significant changes are all aimed at device safety and effectiveness. Medical device manufacturers need to improve their capabilities as well in order to be competitive in the future. Capabilities for the medical device manufacturer are discussed in detail in Chapter 8, Capabilities and MEDICS for an Effective QMS. Regulators expect compliance, customers demand quality, and your business success depend on an efficient and effective QMS.

It is not necessary to change. Survival is not mandatory.

W. Edwards Deming

Establish and maintain

The Quality System Regulation (QSR) says that medical device manufacturers must "establish and maintain" a Quality System. In fact, almost every section of the QSR starts with "establish and maintain." Many people quickly move over the word "establish" without understanding the regulatory significance. Establish means that the medical device manufacturer must:

- Identify the applicable regulations.
- Translate them into appropriate processes documented in written policies, procedures, work instructions, and other documents.
- Implement procedures by training personnel and putting the procedures to use in a controlled manner.
- Create and maintain applicable records to show that these policies and procedures have been *consistently* followed.
- Maintain the procedures to keep them current, make continuous improvements, and if necessary, take corrective action.

So, "establish and maintain" has a broad meaning. Medical device manufacturers cannot simply copy the regulations verbatim and tell employees to follow them. Rather, they must translate the regulations into company processes and procedures that are specific to the manufacturer's products, organization and structure, facility locations, etc. Procedures must be well-written to provide adequate detail and ensure they are understood and consistently followed.

Identifying the applicable regulations is the first step and depends on where you design, manufacture, distribute, sell, install, and service your products. And it depends on the type of products, services, technologies, and activities that you provide. Many medical device manufacturers have regulatory departments to help with this task, interface with regulators, and make regulatory submissions. You'll need to identify the applicable regulatory product codes and product classification for all intended markets. The FDA (*Food and Drug Administration*) has defined classifications for approximately 1700 types of devices and grouped them into 16 medical specialties referred to as classification panels. These are identified in the Code of Federal Regulations (CFR). Examples of these panels include cardiovascular, dental, hematology, and orthopedic products. Each of the devices in these panels is assigned to one of three regulatory classifications. These classifications determine the level of control and premarketing submission required. The device classifications are risk based and depend on the intended use of the device and

Medical Device Quality Management Systems. DOI: https://doi.org/10.1016/B978-0-12-814221-9.00003-8

indications for use. You can find and search these panels on the FDA website under "Classify Your Medical Device."

In this book we will focus on the QSRs of the FDA as well the international standards, ISO 13485 and ISO 14971 which are applicable in many other markets. Of course, there are many other regulations, but these are the places to start creating an effective QMS for most US medical device manufacturers.

Additionally, the FDA does not prescribe in detail how a manufacturer should implement a QMS. Because the QSR applies to many different types of devices, it only provides a "framework that all manufacturers must follow." The QSR requires that manufacturers develop and follow written procedures and fill in the details on how to do things. The regulation requires that "each manufacturer shall establish and maintain a quality system that is appropriate for the specific medical device(s) designed or manufactured, and that meets the requirements of this part." What is "appropriate" depends on the type and classification of your devices, the applicable regulations, and your organization's structure and complexity.

An effective quality management system includes not only the processes and documented procedures, but also the organizational structure and personnel needed to implement those processes and documented procedures. It includes the IT (information technology) infrastructure necessary to perform processes, maintain and control records, and ensure compliance. It includes personnel with appropriate skills, experience, and training to properly and consistently execute processes and procedure. An effective QMS also requires management to express and demonstrate a commitment to quality and compliance. Management must be made aware, via management review, of the health of the QMS. They must provide adequate, trained personnel to ensure a suitable QMS. And they must make corrections and improvements as required. Of course, appropriate records are required to document activities.

FDA inspections are consistent with the requirements to "establish and maintain" procedures. The FDA trains their investigators to the Quality System Inspection Technique (QSIT) in the *Guide to Inspection of Quality Systems*. No matter the type of inspection, the first QSIT step is always to verify that the firm has "procedures that address the requirements of the regulation." You must ensure that your procedures address all the requirements with adequate definition and detail. This seems like a very obvious point, but there are many, many warning letters that indicate how many medical device companies fail to do this.

TIP

For every process, identify and quantify the details of the regulations. As a self-check, calculate what percentage of the requirements are contained in your procedures. In my consulting work, I routinely find that only 40–50% or the regulatory requirements are detailed in procedures. For example, the requirements for Corrective and Preventive Action in 21 CFR 820.100 includes seven specific requirements (Analyze, Investigate, Identify actions, Verify or validate actions, Implement and record changes, Disseminate information, and Submit relevant information for Management Review). It is surprising how many companies fail to include all of these specific requirements in their procedures with the appropriate level of detail. Then we see warning letter citations such as "Your firm's CAPA procedure does not include requirements for verifying or validating, corrective and preventive actions prior to implementation."

Of course, the medical device industry is highly regulated to ensure that customers receive products that are not only safe, but effective. In the medical device industry, people's lives depend on the quality of the products that are made. The regulations have evolved and changed in response to specific problems, health scares, and abuses and violations of the regulations. Additionally, as products become more complex, with new technologies, in a more connected world, we can expect continued change and increasing expectations from regulators and consumers.

It is important that medical device manufacturers have mechanisms to monitor their external environment to understand what issues, changes, and new interpretations are occurring. Regulators do not easily or frequently change the regulations. But, they do express their focus areas, new interpretations, and ideas by:

- Form 483s or inspection reports
- Warning Letters
- Conferences with industry
- FDA "guidance documents" reflect the agency's current thinking on a topic and should be viewed as strong recommendations. They differ from regulation as they are nonbinding (not legally enforceable). Although nonbinding, consider them as valuable information.
- Industry or trade organizations such as Advamed (Advanced Medical Technology Association), AAMI (Association for Advancement of Medical Instrumentation), and MDIC (Medical Device Innovation Consortium).

TIP

Make sure you have a method or a person in charge of monitoring the external environment for changes in regulations, guidance, notifications, and news. Manufacturers should also monitor quality problems, recalls, and MDRs (Medical Device Reports) for competitors, key suppliers, and other manufacturers. By monitoring these sources, companies can continue to develop their awareness and understanding of product issues, uses/misuses of product, and other trends. Knowing about issues due to common technology, suppliers, and materials, etc. can help you to avoid these problems.

So, if a key goal of a QMS is to ensure quality, we must first understand the definition of quality with respect to medical devices. Many organizations define quality as "conformance to specification." This is a very useful definition when making decisions about product acceptance, non-conforming material disposition, servicing, etc. However, it assumes that products meeting specifications meet all customer needs and expectations. Customer needs may include many characteristics not found in product specifications such as customer servicing, training, and support. Additionally, it is important that customer needs be properly identified and documented.

Other definitions commonly used:

- A distinctive attribute or characteristic
- Fitness for purpose

- Features of product which meet customer needs and therefore provide customer satisfaction
- Freedom from deficiencies
- "I know it when I see it."

The QSR defines quality as "the totality of features and characteristics that bear on the ability of a device to satisfy fitness for use, including safety and performance." This is a useful definition but does not explicitly mention meeting customer needs. In fact, the word "customer" is not used at all in the QSR. Even the words "user" and "use" are mentioned sparingly and only in the context of design input and design validation.

By contrast, ISO 13485 makes frequent use of the word "customer." Section 5.2 Customer Focus specifically requires that "top management shall ensure that customer requirements and applicable regulatory requirements are determined and met." Revised in 2016, ISO 13485 is a bit more advanced in adding an explicit focus on the customer.

Juran's Quality Handbook makes a distinction between the "little q" and the "big Q," encouraging broader thinking about quality. For example, products should be thought of more broadly as all products, services, and information. So, we must think not only of product quality but service quality such as providing training, responding to questions, and managing complaints. Quality is not only a factor in customer satisfaction, it is a factor in business success. Evaluation of quality should not be limited to conformance to specification but also seen as satisfying customer needs.

So, each medical device manufacturer should describe their own definition of quality. They should interpret how customer expectations and needs fit into their vision of quality and compliance. Use the concept of conformance to specification for acceptance criteria, dispositioning nonconforming materials, and monitoring process performance. But, think a bit more broadly when innovating, designing new products, and thinking about customer needs.

Medical device manufacturers need to have key capabilities that are vital for creating a healthy and robust QMS. Without these capabilities, they cannot thrive, or even survive, in a complex, competitive, and ever-changing industry. These key capabilities, or Quality System **MEDICS**, are:

- Monitor—Ability to measure and monitor the health of the QMS and processes
- Embrace—Ability of companies to embrace a culture of quality and compliance
- Define—Ability to understand risks, prioritize issues, and define needed improvements
- Identify—Ability to self-identify problems
- CAPA and Improvement—Ability to fix problems robustly, completely, and sustainably

- **S**hare and Communicate—Ability to share and communicate key information in a transparent manner.

Although these capabilities seem simple and self-explanatory, they are not easy for companies to achieve in a deep and meaningful manner. Even small weaknesses in these areas can create serious problems. Medical device companies should self-evaluate their capabilities to help create their strategy and plans for improvement. The Quality System **MEDICS** will be discussed in greater detail in Chapter 9, Compliance Must Result in Improved Quality.

The cost of ineffectiveness can be enormous. An ineffective QMS may be characterized by quality problems throughout the product lifecycle:

- Nonconforming material
- Scrap
- Rework
- Complaints
- Medical Device Reports (MDR)
- Recalls
- Customer dissatisfaction
- Declining sales and market share

An ineffective QMS may also be characterized by compliance issues such as:

- Internal audit observations
- Repeat observations
- Corrective and Preventive Actions (CAPA)
- Notifications of issues from external organizations such as the FDA and Notified Bodies
- Warning letters and Consent Decrees
- Public visibility to warning letters and consent decrees resulting in loss of customer trust and good will.

To simplify, the concepts of "establish and maintain" mean that medical device companies need to:

Say what you do, do what you say, and prove it!

Attention to detail and consistent execution are the keys here. As Warren Buffet says, "It is not necessary to do extraordinary things to get extraordinary results." A quality system ensures consistent execution of everyday processes to achieve extraordinary quality and compliance results. In a medical device quality management system, you must sweat the small stuff!

Common problems with "establish and maintain" include:

- Failure to correctly identify the applicable regulations
- Failure to translate <u>all</u> requirements from the regulations into documented procedures

- Failure to provide adequate detail on <u>how</u> to perform the process including roles and responsibilities, process steps, required equipment or gages, etc
- Failure to train personnel and implement procedures
- Failure to provide adequate documentation, or proof, of consistently following procedures.

WARNING

When mergers and acquisitions are made, management frequently wants to leave the acquisition alone without any "bureaucracy." This can leave you with a big blind spot. Your integration strategy should include making sure they have a fully established QMS with all the associated policies, procedures, infrastructure, organization, and skilled and trained resources.

"Say what you do." An effective QMS depends on well-written, clear, concise policies, procedures, and work instructions. It is easy to under estimate the importance of well written procedures. But, well written procedures can make or break your Quality Management System. Writing good procedures is a capability that every company must have. And it takes certain skills. However, these skills that can be learned. Yet, procedures are often written by personnel who have understanding of the subject matter but are not skilled at translating that knowledge into written procedures that are easy to understand and consistently follow. This results in procedures that are vague, unclear, incorrect, or difficult to follow. And that means problems.

Many times, personnel naively think that writing procedures is only about the technical or subject matter content. This is incorrect thinking. Procedures are the very means by which you communicate expectations to employees. And well written procedures are easier to understand and follow. That means consistency in following procedures as well as the resulting output of processes. Consistent output leads to consistent product quality and services resulting in improved customer satisfaction. And that's good for your business. Writing excellent standard operating procedures (SOPs) is covered in more detail in Chapter 4, QMS Structure along with QMS structure.

"Do what you say" means that you must properly implement procedures including making sure that personnel understand and are trained to procedures. A good training program is an important part of an effective QMS. The regulations per 21 CFR 820.20(a) say that management with executive responsibility "shall ensure that the quality policy is understood, implemented, and maintained at all levels of the organization." Further, per 21 CFR 820(b)(2), "each manufacturer shall provide adequate resources, including the assignment *of trained personnel*" to meet requirements. Additionally, 21 CFR 820.25(b) requires that each manufacturer "shall establish procedures for identifying training needs and ensure that all personnel are trained to adequately perform their assigned responsibilities. Training shall be documented." And there are additional requirements to ensure that personnel understand the significance of their work and are "made aware of device defects which may occur from the improper performance of their specific jobs."

There is a difference between education and training. Education is the process of acquiring knowledge and information, usually in a formal manner. Education enables learners to acquire new knowledge and to think. Education will expand an employee's knowledge base. By contrast, training is a process of acquiring proficiency in some skill or a specific process. Training will enable an individual employee to understand a required task or follow a specific procedure. For example, according to American Society for Quality, a certified quality engineer (CQE) has demonstrated proficiency in the principles of product and service quality evaluation and control. This body of knowledge includes, but is not limited to, "development and operation of quality control systems, application and analysis of testing and inspection procedures, the ability to use metrology and statistical methods to diagnose and correct improper quality control practices, an understanding of human factors and motivation, familiarity with quality costs concepts and techniques, and the knowledge to develop and administer management information systems and to audit quality systems for deficiency identification and correction." This is obviously a very valuable body of knowledge. But the CQE must still be trained to follow company specific procedures for acceptance activities, validation, creating sampling plans, etc. The CQE's education enables the correct implementation of company procedures.

Every employee from the CEO to the manufacturing representative to hourly employees on the manufacturing floor must have the necessary education, background, training, and experience to assure that all activities are correctly performed. This is demonstrated by accurate job descriptions that describe the activities and duties that are to be performed. The job description should define the required education and experience. Each job description must also have an associated training plan that describes training needs including all the policies and procedures that the job requires. In some cases, this can become an extensive list of policies, procedures, and work instructions. Compliance to the training plan should be reviewed annually.

> **TIP**
>
> Annual review of training can easily be connected to annual performance reviews.

It is important that management with executive responsibility not see these requirements as a burden, but as a cost-justified investment. This should result in necessary support and commitment to a comprehensive training program. Training is an important part of creating a culture of quality, employee commitment, and a focus on prevention.

An effective and efficient QMS includes a new hire training program that provides an introduction to the highly regulated medical device industry and helps employees understand the importance and relevance of their roles. This training should introduce employees to company vision, quality policy and objectives, and the basics of a quality system as defined by the quality manual. All personnel should, at a minimum, be trained on the quality policy, good documentation practices (GDPs), training requirements and procedures, and responsibilities to promptly report alleged complaints to the complaint handling unit.

WARNING

One of the most fundamental compliance mistakes made is failure to follow your own internal procedures. You must ensure all personnel follow your internal procedures! Always! Failure to follow written procedures is a very common but very serious problem within your Quality Management System. Failure to follow procedures is not only a compliance issue, it can lead to severe quality problems. All failures to follow procedures should be taken seriously and investigated. CAPAs frequently and superficially identify the root cause of issues as "lack of training" with a corrective action of "retrain the operator." But, if this happens more than one or two times, you must ask why you have so many training failures. What is wrong with your training program that allows this to happen? Why doesn't management ensure they have an effective training program? Use the "5-why's" technique for root cause analysis covered in Chapter 15, Alphabet Soup.

Consider techniques for evaluating training effectiveness. Most companies have varied means of training such as:

- Self-learning by reading a procedure. Effectiveness is demonstrated by signing an acknowledgment that the training is complete. This may be useful for simple, easy to understand procedures and updates to procedures.
- Training may be accomplished via online training modules that employees complete at their convenience. Effectiveness may be demonstrated by completing a short quiz online.
- Manufacturing operators may receive personalized training on a manufacturing process. Effectiveness is demonstrated by having work output inspected for defects and workmanship for a prescribed period of time.
- An engineer may attend a training course on validation and use that training to demonstrate applicable education.
- More complicated processes may require extensive on the job training and supervision until an employee is formally certified.

Every medical device manufacturer must establish a comprehensive training program taking into account the needs and opportunities of all personnel. The company should determine the structure, approaches and mechanisms for delivering and controlling training for all levels of employees. This includes content, delivery, and determining the effectiveness of training. Long-term training effectiveness may be evaluated by monitoring process outputs and process performance, and individual performance.

Training is not a one-time event. Training may be reinforced by management, coaching, positive and negative reinforcement. Training is an on-going process and an investment in prevention.

Care should be exercised in instances of personnel failing to follow procedure, especially when using retraining as a corrective action for a CAPA. When retraining is used as a corrective action in a CAPA, it is important to demonstrate the effectiveness of the training. This can include short-term effectiveness such as verifying that appropriate personnel have completed the training and have understood and adopted it. But, think more broadly. If one person needs to be retrained, are there others? Be sure to think of all appropriate personnel who need to be retrained. Also, use the 5-Why (see Chapter 15: Alphabet Soup for more detail) technique to dig deeper. Why did the performer fail to do their job? Was the procedure unclear or incorrect? Why? Was the original training ineffective? Why? Were the original training materials incorrect or incomplete? Why?

Because training and awareness declines with the passage of time, it is important to repeat and reinforce critical information such as the quality policy. For example, all employees should receive annual formal refresher training on the quality policy. The quality policy may be reinforced by putting the quality policy on posters throughout your facilities or on computer logon screens.

The use of IT systems and technology is not required to control and maintain training records. But, it can greatly improve the training process and enable managing vast amounts of training documents in an effective and efficient manner. Some integrated quality management systems can greatly mistake-proof training issues. For example, an employee cannot login into or use a particular machine without required training being complete and documented. Another feature is that IT systems autogenerate notification of updates to documents, training requirements, and reminders for all personnel that have a specific procedure on their training plan. Manufacturing systems will not allow an employee to perform the process if they do not have documented, effective training on the revised procedure. After the implementation date of the change, an operator will no longer be able to use the old revision in batch records.

Training and education are important tools for implementing an effective and efficient quality management system. A strong training program, aligned with quality objectives and personnel development, is a powerful tool for preventing quality and compliance problems. A suitable quality management system is dependent on the knowledge, skills, and training of the workforce. The status of the training program and results should be communicated regularly during management review.

"Do what you say" also depends on proper version control and change management to ensure procedures are the correct and current version of procedures are followed. Change control is an enabling process for quality and compliance. Clear and controlled linkages to CAPA and continuous improvement are essential.

"Prove It" means that you need to maintain and control a vast amount of documentation including the records that show personnel followed procedures consistently. And yes, 21 CFR 820.40 has specific requirements for the control of documentation. Medical device manufacturers shall establish and maintain procedures for:

- Document approval and distribution
  - Designated individual(s) must review documents for adequacy and approve prior to issuance. This sets a basic expectation of good quality work and management responsibility to ensure that happens consistently. Designated individuals can include management, subject matter experts, and stakeholders that must execute to the written procedures, as well as personnel in quality or regulatory positions. You must determine what is suitable for your company structure, complexity, and risk level. To provide proof, there are requirements for signature and date of the approver(s).

WARNING

The regulatory requirement is not just to approve documents. It is to "review for adequacy and approve." All approvers have serious responsibilities to really review the document for adequacy before approving. It must not be a rubber stamp.

  - There are also expectations that the documents are available at "all locations for which they are designated, used, or otherwise necessary." This simple requirement is often violated. I've found many instances where manufacturing procedures were not available in the manufacturing area, but were locked in a supervisor's desk, or were written in a language that manufacturing personnel (obviously) could not read. They must be readily available to the users and must be written in a manner that can be understood by them.
  - All "obsolete documents shall be promptly removed from all points of use or otherwise prevented from unintended use." This is a key concept in ensuring the correct version of the document is used. For companies using paper based systems, it is more difficult to control and identify official documents for use. Unofficial documents should be identified as such and used for reference only and not to actually follow them. Personnel that are actually using the procedure must use an official version only. Companies using electronic document management systems (EDMSs) have an easier time of controlling and obsoleting documents. Some EDMSs also enhance training and implementation of revisions to documents.
- Document changes shall be reviewed and approved by an individual(s) in the same function or organization that performed the original review and approval. This ensures that there are no unapproved changes to remove or change important requirements from a document. For example, someone in the manufacturing organization could not remove a key requirement without

review and approval from the quality organization. Unfortunately, this is one of those areas where the regulations were based on previous compliance issues. The use of a RACI (Responsible, Accountable, Consulted, Informed) Model is helpful for defining owners, contributors, and approvers of documents. See Chapter 15, Alphabet Soup for detail.

- The change will include identification of the change, affected document(s), and effective date.
- When changes are made, they shall be communicated to the appropriate personnel in a timely manner. This is another case where previous transgressions have resulted in a regulatory requirement. Pay attention to timeliness.
- Maintain records of changes to documents including a description of the change, identification of the affected documents, the signature of the approving individual(s), the approval date, and the date the change becomes effective. The use of a revision history at the end of each document is necessary to accomplish this along with an overall document and change control process.

Throughout the regulations, we see additional requirements such as:

- Maintain records
- Document approvals
- Data shall be approved in accordance with 21 CFR 820.40
- Results and conclusions shall be maintained
- Outputs shall be documented.

"Shall be documented" is frequently the last sentence of regulatory requirements. ISO 13485:2016 uses similar language with most clauses ending in "records shall be maintained." These requirements set the expectations, over and over again, that controlled, written records are a fundamental expectation for an effective Quality Management System. Yes, it is a lot of work to create, control, and maintain all that documentation. But, the expectation is reasonable given that we are talking about making medical devices that can impact people's lives, health, and safety. Due to the large volume of documents and records every medical device manufacturer must manage, you also need to be very efficient and effective at document and records control! A thorough, methodical approach is necessary to maintain good records. In the case of quality problems, proper change control, revision history, and training records will be critical to "bounding" of affected product.

Records provide the proof that you followed the requirements of your own quality management system. There are also general requirements for records in 21 CFR 820.180:

- All records shall be maintained at the manufacturing establishment or other location that is reasonably accessible. This allows companies to store old hard-copy documents offsite at a facility such as Iron Mountain, but the expectation is that it is reasonably accessible. During an FDA inspection, you want to make sure that these documents can be found and delivered to you in a day or so. Companies that use electronic data management systems do not have to worry about storing paper documents offsite.

- Records must be made available for review and copying by the FDA. In practice, the FDA frequently asks for written documents. But, they can also request copies to take with them. Sometimes they look live at your electronic systems or request flash drive (or other storage media) with information on it. This will be discussed more in the section on inspection management in Chapter 16, FDA Inspection Readiness.
- Records shall be stored to minimize deterioration and prevent loss. Preservation of documents is required (either hard or electronic) copies. For hard copy documents, you need to worry about damage from fire, water, insects, and rodents. For electronic documents, you need to worry about Part 11 compliance, back-up, hacking, security, and data integrity. Consider planned destruction of obsolete information. Incorporate legal holds as necessary.
- Identify a disaster recovery plan. There are numerous medical device manufacturers in Puerto Rico that had to implement their disaster recovery plans after Hurricane Maria devastated the island in 2017.
- Records that are deemed confidential may be marked to aid the FDA in determining whether information may be disclosed under public information regulations. Note: In warning letters made available to the public, confidential information is redacted and replaced by "(b)(4)," a statute of the Freedom of Information Act (FOIA) that protects disclosure of trade secrets or financial information where disclosure is likely to cause harm.
- Records shall be retained for a period of time equivalent to the design and expected life of the device, but in no case, less than 2 years from the date of release for commercial distribution. Every medical device company must have a clear records retention procedure. There may be other reasons to retain documents as well such as legal needs, and environmental or OHSA regulations, so be sure to work with your legal department on this.
- Exceptions include the content of management review, quality audits, and supplier audit reports that do not need to be shown to the FDA. Obviously, many manufacturers are unwilling to feed information about known quality system problems to the FDA. However, you must show records that the activities have been done. For example, you do need to show that management review was conducted although you do not need to show the details of the information that was reviewed. And you are also required to show that any required corrective actions have been undertaken.

The concept of "Prove it" also requires good documentation practices (GDP). All employees are responsible for GDPs. All documents must be clear and legible. GDP standards include:

- Notes must be contemporaneous with the events they describe.
- All records must be complete and accurate.
- Records must be free from errors.

- Documents must be approved, signed, and dated by authorized personnel only.
- Signatures provide evidence that procedures are followed.
- Signature pads, scanned signatures, or duplicated signatures may not be used.
- Documents must be signed in permanent ink.
- Adequate space must be left for hand written entries.
- Hand written entries should be made with permanent ink.
- No spaces for hand written entries are left blank (if unused, they may be lined out or indicated NA). This prevents unacceptable future additions.
- Ditto marks are unacceptable.
- Documents are retained per records retention requirements.
- Documents are maintained in acceptable condition. Paper documents should be stored to prevent fire, water, pest, or other damage. Electronic records must be maintained to show all change history, prevent unacceptable alteration, and have back-ups.
- Electronic data management systems must be validated.
- Electronic records are backed up to prevent losing records.
- Document modifications must be controlled:
  - Hand written changes must be signed and dated;
  - The reason for change must be indicated (e.g., "entry error" or "incorrect date").
  - Use only a single line through information that needs to be corrected. Do not obliterate previous information. Changes must be signed and dated.
  - Do NOT scribble out information, use white out, or write over previous data.
  - Do not put sticky notes, crib notes, cheat sheets, etc. in documents
  - Only authorized individuals may make changes to controlled documents
- Changes to procedures must go through full document control.
- Controls must exist to prevent use of superseded documents.
- Electronic records can only be modified by authorized personnel (controlled by password or other means).
- History of changes, additions, and deletions must be maintained.
- Supervisors, managers, and all approvers are trained and appropriately authorized to review and approve documents that they do approve.
- An approved list of originators, designees, and reviewers is maintained.
- Designees add the words "signing for" and the original approver's name.
- Designees must also have the knowledge and skills necessary for the activity. Attach a copy of the delegation letter.
- Document retention requirements must be identified and documented.
- Understand and control different copy types, such as:
  - Master copy
  - Controlled copy
  - Uncontrolled copy
  - Superseded copy
  - Obsolete copy

- Use legal dates and formats including month, day, and year. Use a consistent prescribed format. Warning: multinational companies must be aware that some countries use DD/MM/YY and others use MM/DD/YY. This can be confusing for something like 02/03/18. Does this mean February 3rd or March 2nd? You'll often see quality professionals use the date format DD Month Year such as 02 Mar 2018 to prevent any chance for confusion. If electronic time stamps are required define format as AM/PM or 24 hours.
- Document changes should have an effective date.
- Originators/authors should always sign documents.
- Never back date documents.
- Controlled documents such as laboratory note books should have consecutive page numbers. Put a single line through unused areas to prevent later additions to data.

FDA investigators are trained and skilled in looking for documentation errors, inconsistencies, and even evidence of fraud (alterations, missing pages, etc.). Make sure you have a good process for documentation practices and ensure that it is followed. As with all processes, there should be a process owner knowledgeable of the regulations. Most companies have a document control function for this. All employees should be trained on GDPs to understand the FDA expectation "If it's not documented, it didn't happen."

Data Integrity is an important consideration in accurately maintaining records. It applies to both paper and electronic records. Data have integrity if they are retained, truthful, accurate, complete, and verifiable. *21 CFR 11 Electronic Records and Electronic Signatures* sets requirements for data integrity in electronic records. IT systems must be validated to ensure that they are capable of meeting their intended purpose. Ensure that validation includes the entire system including hardware, software, devices, operators, and procedures.

TIP

Don't relegate data integrity compliance to the IT department alone. Although they have technical expertise, they are not the process owners for the quality management system and associated processes. Data integrity requires a team effort from all stakeholders. Establish a data governance process to ensure that accurate and complete GMP data is appropriately viewed as a valuable corporate asset. Include Quality, Regulatory, IT, Process Owners, and Data specialists on your board.

In conclusion, every medical device company needs to have vision of quality to establish and maintain a suitable quality management system. It starts with identifying the regulations that are applicable based on the products, risk levels, and locations where you design, manufacture, distribute, and sell products and services. These regulations must be accurately translated into clear, complete, and accurate policies and procedures. In this chapter, we explored the expectations for "establish and maintain" and what that means in practical terms of say what you do, do what you say, and prove it. The concepts are easy. It is consistent and accurate execution that is more demanding.

QMS structure

4

Processes and procedures—what is the difference? The introduction to ISO 13485 provides a good description regarding a process approach and how that differs from procedures. It describes a process as "any activity that receives input and converts it to output." Often the output of one process becomes the input of another process. By contrast, a procedure is an established, standardized, and *documented* way of conducting a process. A procedure is the written document that describes the details of the process.

Further, ISO 13485:2016 0.3 indicates that:

For an organization to function properly, it needs to manage numerous linked processes. The application of a system of processes within an organization, together with the identification and interactions of these processes, and their management to produce the desired outcome, can be referred to as the process approach. When used within a quality management system, such an approach emphasizes the importance of:

1. *Understanding and meeting requirements;*
2. *Considering processes in terms of added value;*
3. *Monitoring results of process performance and effectiveness;*
4. *Improving processes based on objective measurement.*

A system is an organized network of processes forming a unified and effective whole. A well-established quality management system (QMS) includes the processes, structure, roles and responsibilities that enable companies to ensure safe and effective product for their customers. A QMS uses a hierarchy of documents (generically known as *procedures*) to describe the structure, responsibilities, and content of processes. Fig. 4.1 describes a typical QMS hierarchy for a company of low to medium complexity. More complex companies/organizational structures may have additional layers such as corporate level policies and procedures.

The *quality policy* provides a basis for establishing quality objectives. It is jointly developed by management and quality experts. It expresses management's views, commitment to, and expectations for quality, compliance, and customer focus. It describes values such as commitment to providing quality products to satisfy customer needs; comply with regulatory requirements; and to continual

Medical Device Quality Management Systems. DOI: https://doi.org/10.1016/B978-0-12-814221-9.00004-X

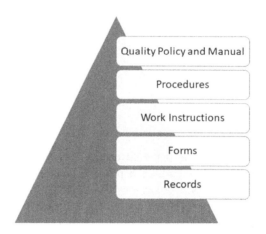

FIGURE 4.1

QMS hierarchy.

improvement. It must be communicated and understood by the organization and periodically reviewed for suitability.

The quality system regulation (QSR) defines the quality policy as "the overall intentions and direction of an organization with respect to quality, as established by management with executive responsibility." ISO 13485:2016 has slightly more detailed requirements in clause 5.3 Quality Policy. "Top management shall ensure that the quality policy:

1. Is applicable to the purpose of the organization;
2. Includes a commitment to comply with requirements and to maintain the effectiveness of the quality management system;
3. Provides a framework for establishing and reviewing quality objectives;
4. Is communicated and understood within the organizations;
5. Is reviewed for continuing suitability."

Additionally, clause 5.4.1 Planning—Quality Objectives requires that

Top management shall ensure that quality objectives, including those needed to meet applicable regulatory requirements and requirements for product are established at relevant functions and levels within the organizations. The quality objectives shall be measurable and consistent with the quality policy.

TIP

It is important that creating a quality policy is not seen as a paper exercise, but as the means by which management communicates the importance of and commitment to quality, compliance, and customer focus. Management can reinforce the importance of the policy by mentioning it (or the

(Continued)

> **TIP (CONTINUED)**
>
> principles in it) consistently during employee meetings, in business reports, and other communications. Demonstrating a firm commitment to quality and compliance is an also an important factor in building customer trust.

For most companies, the quality policy is a relatively short document, sometimes only a few paragraphs long. It should be highly visible and well communicated. It should be on the training plan of every employee in your company. The quality policy should be contained in your quality manual and be highly visible throughout the facility. It can be made more visible by putting it on posters throughout the facility, on computer-login screens, on TV monitors in the cafeteria, and in annual or business reports. Management with executive responsibility should regularly emphasize the importance of the quality policy during employee meetings, business updates, etc. Routine communication, visibility, and management emphasis on the quality policy can have a significant positive impact on the culture of quality and compliance within your company.

Expect Food and Drug Administration (FDA) investigators to examine if your employees know and understand the quality policy. During FDA inspections, the investigators routinely ask employees if and how they know the quality policy. This is another reason that the quality policy should be highly visible and regularly emphasized by management.

The *quality manual* is the next document in the quality system hierarchy. The quality manual describes the minimum requirements for the structure, ownership, and content of the quality system. The quality manual should show how the pieces work together as one unified whole. The structure and content of the quality manual will depend on the size, structure, and complexity of your organization. The quality manual should describe or contain a diagram of your major processes and their relationships. It may include processes that are necessary for product realization throughout the product life cycle such as:

- Design products
- Produce products
- Supplier and purchasing control
- Installation and servicing activities
- Identification and traceability
- Clinical studies and product performance data

It may also include enabling processes, which are not specific to any particular product, such as:

- Management responsibility
- Quality audit
- Personnel/training
- Document controls

- Corrective and preventive action (CAPA);
- Records management
- Statistical techniques

Below the quality manual are *standard operating procedures (SOPs)*. Procedures describe the overall process, ownership, and actions required. The SOPs should define the requirements from the regulation and describe *what* must be done.

Below SOPs are *work instructions* (or WIs). WIs are similar to SOPs but are characterized by increased focus and further detail on *how* to perform a task. A work instruction describes how to undertake a specific part of a function of activity. A simple process may be adequately described in a single SOP. More complex processes, such as design control may require several SOPs and subordinate WIs (e.g., design planning WI, design review WI, and design transfer WI, etc.) to provide adequate detail.

WARNING

Do not copy the regulations word for word! It is important to interpret them and apply them to your company's unique structure, products, and business situation. Your quality system needs to be **appropriate** for the types and risks of the devices that you make. It needs to provide sufficient detail on how to perform processes.

Below WIs are *records*. Records are a way of documenting that policies, procedures, and WIs have been followed. Records may be forms that are filled out, a stamp of approval on a document, a data entry in the manufacturing system, or a signature and date on a document such as a routing sheet. Records are used to provide traceability of actions taken on products or processes.

Sometimes companies also use "guidance documents" but these can be problematic as some personnel see them as a suggestion only and feel they have permission not to follow them. Frequently, when guidance documents are perceived to add work, they are seen as unnecessary and burdensome. It is like using the word "should" in procedures and it is usually ignored. Use caution when creating guidance documents and using the word "should" in procedures.

The quality system means the totality of organizational structure, responsibilities, procedures, processes, and resources for implementing quality management. Your quality manual should include key information on process ownership, responsibilities, roles and responsibilities. Organizational charts, job descriptions, and training records are additional mechanisms used to define and detail the overall quality system. As such, FDA investigators frequently ask for these documents.

For companies that are just getting started, consider the following priorities in establishing your QMS:

- Set the foundation and structure by understanding your business objectives, product types, and regulations.
- Appoint a management representative.
- Start by creating the basics—a quality policy, objectives, manual.
- Consider people—define roles and responsibilities, job specifications, authorities, training requirements, process ownership.
- Then add foundational enabling processes such as document control, internal audit, and CAPA that interreact with and control all other processes.
- Start regular management reviews to determine status and plans for meeting quality objectives.
- Determine essential processes necessary for your immediate business objectives and stage in the product lifecycle:
  - Product development requires:
    - Design control
    - Risk management
    - Supplier management
    - Lab controls
  - Manufacturing transfer requires:
    - Production and process controls
    - Validation
    - Labeling and packaging control
    - Acceptance activities
    - Acceptance status
    - Identification and traceability
    - Change management
    - Nonconforming material
    - Statistical techniques
  - Release to market requires:
    - Installation
    - Servicing
    - Handling
    - Storage
    - Distribution
  - Post market activities require:
    - Complaint management
    - Adverse event reporting
    - Corrections and removals
    - Customer feedback

Documentation, Documentation, Documentation! Like the well-known realtor's slogan, it is all about documentation. Documentation is critical to an efficient and effective QMS. Regulators will request records as proof that you have met regulatory requirements. If activities were not documented, you cannot prove they were done.

You must have a strategy for control and maintenance of documentation that includes the following types of records:

- Device Master Records per 21 CFR 820.181
- Device History Records per 21 CFR820.184
- Design History Files per 21 CFR 820.30(j)
- Quality System Records per 21 CFR 820.186
- Complaint Files per 21 CFR 820.198
- Purchasing data per 21CFR 820.50(b)

Although these are the types explicitly mentioned in Subpart M of the QSR, there are many, many other types of records as well such as training, calibration, service, installation, returned goods, acceptance activities, environmental controls, audit, management review, supplier controls, nonconformances, and change orders. CAPA records are of particular importance. Determine your processes, structure, organization, and control of all of these records.

A Design History File (DHF) per 21 CFR 820.10(e) means a compilation of records, which describes the design history of a finished device. Per 21 CFR 820.30(j), it contains or references all records necessary to establish compliance with the design plan and regulations including design control procedures. The DHF is not only a regulatory requirement but also provides significant value to the manufacturer in understanding development successes and failures, and approaches for new designs. It contains valuable verification and validation protocols that are not in the DMR. Despite the name "file," the documentation does not need to be housed in one continuous filing cabinet. Rather, the compilation of applicable documents can be housed in various locations or an electronic data management system (EDMS), but it must be readily available. Typical documents include:

- Design plans
- Design review meeting minutes
- Drawings, specifications, assembly drawings
- Procedures
- Risk analysis
- Engineering notebooks
- Component qualification
- Biocompatibility
- Verification protocols and reports
- Validation protocols and reports
- Design transfer checklists

The DHF is a living document and must be maintained throughout the product life cycle. Thus, design changes must be properly documented and added to the DHF.

The total design output includes the device, its packaging and labeling, as well as the DMR that includes device specifications and drawings, as well as the instructions and procedures for production, installation, maintenance, and servicing. The DHF is the basis or starting point, for the Device Master Record (DMR). The DHF differs from the DMR as it illustrates the history of the device and is necessary to show that manufacturers exercise control over and are accountable for the design process.

The DMR per 21 CFR 820.3(j) means a compilation of records containing the procedures and specifications for a finished device. The DMR contains or references the current information necessary to manufacture, install, maintain, and service the device. The DMR refers to the total record and any of its individual records. The DMR for each type of device shall include or refer to the location of:

1. Device specifications
2. Production processes
3. Quality assurance procedures and specifications including acceptance criteria
4. Packaging and labeling specifications
5. Installation, maintenance, and servicing procedures

The Device History Record (DHR) contains information for each lot, batch, or unit of product to demonstrate that the device was manufactured in accordance with the DMR. The DHR shall include or refer to the location of:

1. Dates of manufacture
2. Quantities manufactured
3. Quantity released for distribution
4. Acceptance records
5. Primary identification label
6. Any unique device identifier or other identification
7. Also materials used

Manufacturers must consider how to manage documents throughout the total product life cycle. This is sometimes referred to as PLM (*Product Lifecycle Management*) and formalizes the processes required for product realization. PLM systems increase the ability to organize, store, manage versions, maintain, search, and control product documents. They increase the ability to collaborate, review, and approve documents, regardless of physical location.

Quality system records are those that are general in nature and not specific to any particular product. They include general documents such as records required for management responsibility.

Determine the types of records you will be generating for your QMS and determine the requirements for them. Determine the owners of the documentation, how long you need to maintain the documentation, how you will control and maintain the documentation and in what form. Figure out where you will keep the documentation. Consider flexibility and scalability for future growth.

> **TIP**
>
> Create a document control governance board. A cross-functional board, including quality systems, information technology, and process owners for the product realization processes can collaborate to create and implement an effective strategy for document control and IT infrastructure. The Document Control Board should evaluate current state and a strategy for improvement and growth.

Electronic data management systems are available and offer significant advantages over paper-based systems. They can be configured to automate your document control processes within the framework of regulatory requirements such as 21 CFR Part 820 and ISO 13485. Although there are investment costs for an EDMS, there are also significant savings in terms of labor, regulatory conformity, maintenance, and improved security. An investment in an EDMS will pay for itself. Obviously, an EDMS will improve the efficiency of your QMS, but it will also improve the effectiveness of your QMS. It can help to more consistently create and control records that conform to requirements.

An EDMS can also facilitate training of employees for new and changed procedures. And it can mistake-proof common problems with employees performing activities without documented training in place. An EDMS system ensures timeliness of approvals, creates reminders and follow-ups for tasks, facilitates archiving and retrieval of documents, and provides back-up for disaster recovery.

> **TIP**
>
> Switch to an EDMS as soon as possible to set a good foundation for future growth and volume of documentation. Conversion and data migration costs and problems will only increase the longer you wait. Sooner is better than later.

Controlled documents require a formal process for authoring, reviewing, approving, publishing, changing, retrieval, and archiving of documents. It assures that:

- Documents are reviewed and approved by quality and other subject matter experts prior to being implemented
- Only the current version of the document is official and available for use
- Unauthorized copies are not used
- Previous or obsolete versions have been removed from use
- Safeguards are in place to prevent unauthorized changes and access
- Archiving and maintenance of documents is managed

Switching to an EDMS should consider preparation, processing, and administration activities. Preparation includes being able to categorize, classify, and tag documents with associated attributes (metadata). Common attributes include title, author, number, version, effective date, etc. They may include other attributes such as applicable site, department, process, or product name. It can be helpful to identify and have consistent formats for various types of documents. A data management system involves the softcopy (e.g., Microsoft Word™) document, its portable document format (PDF) rendering, metadata, and workflow information.

Processing includes authoring, formatting, and reviewing documents to ensure good content. A collaborative environment for sharing drafts, making comments, editing, and approving final documents is helpful and improves efficiency. Approval, change control, security, and distribution of documents need to be done in compliance with 21 CFR Part 11. An effective system gives thought to the end-user needs and how to distribute documents. The process must define rules for how to retire, archive, and dispose of documents in a controlled manner.

An EDMS system also requires administrative activities including access rights. Assign access rights for authors, approvers, general users, and administrators of documents. As with every process there should be metrics to monitor, analyze, and improve the performance of the process.

An effective QMS also requires effective management of changes. Change control covers all changes and alterations of an existing approved document including design, software, labeling, packaging, raw materials, components, manufacturing processes, production equipment, quality system documents, and procedures. Changes may be due to quality problems, marketing or customer needs, regulatory or standard changes, productivity and cost improvements, supplier changes, document or format improvements, continuous improvement, and CAPA. In short, change happens. An effective and efficient change management process is required.

The change control process should provide a structured and consistent approach to:

- Document the details of the change
- Ensure appropriate approvals (including originator functions)
- Document required verification or validation prior to implementation
- Provide change history and retrieval information
- Track changes and providing an audit trail

Change details should include a description of the change, effective date, responsibilities and approvals, revision level, validation, communication requirements, and regulatory submissions. Specific regulations apply for production and process changes (21 CFR 820.70(b)) and design changes (21 CFR 820.30(i)).

TIP

Consider a change control board. Complaints, Medical Device Reports (MDRs), and other quality problems can drive significant changes. Changes to address quality issues have a high likelihood of being inspected by regulators and deserve appropriate attention. Define expectations for identification, documentation, verification and validation, review, and approval of design changes. CAPA files should document the progress and investigation of quality issues while referencing the change control file. Determine who has control and ownership of these types of changes (including quality, regulatory, R&D, and operations).

Records, documents, and change management are one of the seven subsystems described in the FDA's *Guide to Inspection of Quality Systems*. But, this subsystem is not directly inspected. The adequacy and consistency of records and documentation will be quickly determined as the investigator looks at the four QSIT (Quality System Inspection Technique) focus areas (see Chapter 16: FDA Inspection Readiness). The FDA will request procedures and records to inspect the four key QSIT areas. The QIST guidance instructs investigators to "determine if the firm has defined and documented the requirements (CAPA, design, etc.) by looking at procedures and policies, and then you will bore down into records, using the sampling tables, where appropriate, looking at raw data to determine if the firm is meeting their own procedures and policies, and if their program for executing the requirement is adequate." Providing consistently acceptable documents, procedures, and records in a timely manner is a key to good inspection results. Document and records control is the lens through which the investigator will see your QMS.

WRITING EXCELLENT SOPS

Many times, people naively think that writing good SOPs is just about the technical or subject matter content. This is a big mistake. Well-written SOPs are extremely important and are the means for successful communication to your employees. They provide standard instructions on what to do and how to do it. They are a mechanism to ensure that procedures are understood and consistently followed. Consistently followed procedures result in more consistent output. And more consistent output means more consistent product quality. That means improved customer satisfaction.

Well-written procedures also send a message to regulatory inspectors that you understand the regulations, are organized, and are in control.

> **TIP**
>
> Create a "procedure profile" for all procedures defining document type (policy, procedure, WI, form), level, owner, applicability (entire corporation, some sites, some manufacturing lines, some products), impact, and training requirements.

Benefits of clear operating procedures:

- Ensures consistent, repeatable processes
- Provides the basis for training
- Facilitates training
- Provides metrics or describes outputs to measure the perfomance of the process
- Provides a baseline for assessment and improvement

- Shows regulators that you have an accurate, compliant quality system.

  Well-written procedures also:

- Show clear linkage to the regulations
- Are easy for users to understand and FOLLOW, which:
  - Drives compliance
  - Reduces variability and inefficiency which is costly
  - Improves quality and consistency, which enhances customer satisfaction
- Identifies risks, hazards, and how to avoid them

Common compliance issues seen in 483 observations and warning letters include:

- Failure to establish and maintain or follow written procedures
- SOP XXX is incomplete
- Failure to develop, maintain, or implement written procedures
- Failure to develop adequate written procedures
- There are no procedures for...
- The procedure does not establish...
- There are no instructions
- Procedure XXX lacked details
- The procedure does not identify who
- The procedure does not establish how
- There is no evidence your firm's procedures have been implemented
- Failure to follow written responsibilities and procedures
- Procedures for XXX have not been established

WRITE IT RIGHT

Writing good SOPs is a skill that can be learned. Good writing skills can prevent these common mistakes:

- Incorrect references
- Nonstandard terminology
- Long, wordy paragraphs
- Out of date information
- Requirements, terminology, and definitions that are inconsistent with higher level documents
- Lack of flexibility (adding too many requirements). For example: "Enter the data into the Dell XPS13 computer on the manufacturing floor." What happens when the Dell computer is replaced by an HP computer?
- Vague terminology. For example: "Conduct Management Review two times annually." Does this mean two times per calendar year, two times in a fiscal year, or two times in the year since the first review?

- Input and output processes are conflicting, making it impossible to comply with both.

Goals for writing good procedures are to make sure they are:

- Clear and concise
- Unambiguous
- Flexible where possible
- Strict where necessary
- Easy to read, using picture, diagrams, flow-charts, etc.
- Apply mistake-proofing concepts
- Use words carefully such as:
  - "Can" indicates a capability or possibility
  - "May" indicates permission
  - "Should" indicates a recommendation but is an ambiguous term and must be used with caution;
  - "Shall" is used to indicate a firm requirement

Steps to writing good procedures:

1. Choose a format for the instructions:
   a. Simple steps (e.g., 1, 2, 3) is good for short routing procedures.
   b. Hierarchical (e.g., 1 a,b,c; 2 a,b,c) is good for long procedures requiring more detail and clarification.
   c. Flowcharts are best for decision making and showing alternate paths.
   d. Procedures with WIs below them are necessary for very involved and complicated processes.
2. Consider your audience. Understand previous knowledge, education, and language skills.
3. Keep your purpose in mind:
   a. Ensure regulatory requirements are met
   b. Ensure safety of the operator
   c. Ensure no impact to the environment
   d. Prevent failures
   e. To be used as a basis for training.
4. Cover the following:
   a. Scope and responsibilities—purpose, limits, and how it is used. Include standards, regulatory requirements, roles and responsibilities, inputs, and outputs.
   b. Clarify terminology—identify acronyms and abbreviations.
   c. Methodology and procedures—list the steps.
   d. Identify needed equipment and supplies.
   e. Provide appropriate health and safety warnings.
   f. Identify cautions and problems.

5. Get input from the experts—the people who will execute the process.
6. Test the procedure—have someone with limited knowledge of the process review it for understanding. Have the people who will execute the process review your final draft.
7. Route the procedure for approvals. This can be manual or electronic depending on the size and maturity of your organization.
8. Implement the procedure. Define formal training requirements. Include evidence that training is effective.
9. Ensure periodic review and, if appropriate, improvement.

Make sure your company procedures all have a consistent look and feel to them. This shows consistency and attention to detail. Use common styles, headers, footers, logos, and formatting schemes. An outline for a typical SOP includes:

- Purpose
- Scope
- References
- Definitions
- Responsibilities
- Related documents
- Process steps—who, what, where, when, how
- Equipment and materials used

Sometimes companies add a background or strategy section, but these are not always necessary. An appendix is sometimes used as well.

All procedures should have headers and footers that include document information such as title, unique document number, author, approvers, revision level, and date. The header may include your company name or logo. The footer should include the page number and number of pages. You may wish to add a routine statement about document confidentiality. There should be a revision history in the document containing the version numbers and general description of the changes.

Warning—Be Concise!

- Use lists and bullets.
- Keep sentences short.
- Use short words and plain language (reference https://www.plainlanguage.gov/ and other sources).
  - Say list—instead of—a list of items.
  - Say warning—instead of—caveat.
  - Say some—instead of—a number of.
- Don't use the word "you"—it is implied.
- Use imperative sentences (like this one).

The paragraph above is much easier to read than the poorly written version below:

A caveat for writing your SOPs: You should write SOPs that are concise, and you should have lists of items and a number of good bullets instead of paragraph format because long wordy paragraphs involve a lot of words and take a long time to read. Also, it is much better for you to use concise sentences instead of using long, wordy, run-on sentences like this one because long, wordy, run-on sentences make it so difficult to follow and read the instructions. And you should be sure to always remember to utilize abbreviated words rather than very lengthy or overly complicated words. For example, you should utilize words like "next to" instead of "adjacent to," or "some" instead of "a number of." You should utilize simple words like "warning" instead of using a fancy word like "caveat" that not everyone immediately knows about. You can reference a number of websites like https://www.plainlanguage.gov/ and others that can help you find shorter words to use. And one more reminder, you don't need to use the word "you" in sentences like this as it is implied. You should really use imperative sentences (unlike this one) because they are so much shorter and so much more precise.

Consistent with the "process approach" described in ISO 13485:2016, consider all the relevant process factors:

- A process is a structured set of activities (process steps) to convert a set of inputs to value added outputs, providing value to customers and stakeholders.
- Inputs are the materials, resources, equipment, tools, data, and information to complete a process.
- Suppliers (or sources) are the providers of the various inputs to your process. They can be considered the internal (prior processes) and external providers (external supplier) of inputs.
- Outputs are the products, services, information, and metrics that result from the process.
- Customers are not only the end customers (users, patients, hospitals) but also the internal users (other processes, functions, etc.) and stakeholders.

Six Sigma methodology commonly uses a tool called a **SIPOC** (**S**upplier, **I**nputs, **P**rocess, **O**utputs, **C**ustomers) to diagram these factors. The SIPOC diagram (see example in Fig. 4.2) can be used to outline these factors as you start to draft your procedures. See Chapter 15, Alphabet Soup for detail. Starting with a good outline makes procedure writing much easier.

Be sure to look at input and output processes to make sure that they are aligned. The output of one process becomes the input to the next process. Often these connecting procedures are written by different people who are not aligned. This is a common source of confusion, creating both inefficiency and ineffectiveness.

A picture paints a thousand words! Use diagrams, flowcharts, process maps, and cross-functional process maps to add clarity and define the order of process steps. Flowcharts can be very useful for documenting decision points and

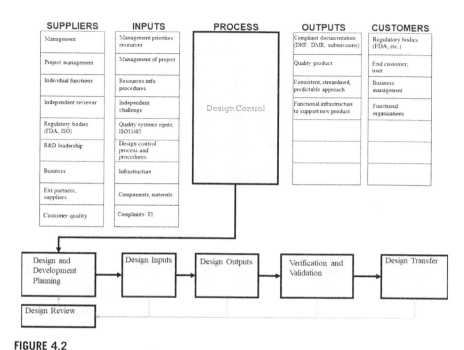

SUPPLIERS	INPUTS	PROCESS	OUTPUTS	CUSTOMERS
Management	Management priorities/ resources		Compliant documentation (DHF, DMR, submissions)	Regulatory bodies (FDA, etc.)
Project management	Management of project		Quality product	End customer, user
Individual functions	Resources/info/ procedures		Consistent, streamlined, predictable approach	Business management
Independent reviewer	Independent challenge	Design Control	Functional infrastructure to support new product	Functional organizations
Regulatory bodies (FDA, ISO)	Quality systems rqmts, ISO13485			
R&D leadership	Design control process and procedures			
Business	Infrastructure			
Ext partners, suppliers	Components, materials			
Customer quality	Complaints/ FI			

Design and Development Planning → Design Inputs → Design Outputs → Verification and Validation → Design Transfer

Design Review

FIGURE 4.2

SIPOC diagram.

alternate steps to take based on the decisions made. For example, a complaint handling procedure may have specific decision trees for particular products and how to determine and document the complaint type and code.

Consider mistake proofing concepts as you write your procedures. Include mistake proofing in all of your procedures. Mistake proofing efforts can include:

- Provide clear, concise, unambiguous instructions.
- Include checklists to ensure required items are completed.
- Use forms to make sure records include appropriate data and information.
- Use automated systems to ensure certain steps are documented or certain information in forms is entered. For example, an electronic batch record cannot be completed if all data are not entered into all required fields.
- Use FMEA (Failure Modes and Effects Analysis) methodology to analyze what process failures can occur, analyze detectability, and severity. Use this to define risks and control measures.
- Use Poka-Yoke concepts to identify errors, improve detectability, and actions (including controls, shutdown, and warnings).
- Add precautions and warnings (if appropriate) to your procedures.

Make sure that the personnel writing procedures are trained in writing good procedures. Otherwise, if you have a lot of procedures to write, hire a technical writer to translate the procedures created by Subject Matter Experts.

Other tips:

- Use precise language.
- Create clear definitions for key terminology and acronyms. Stay consistent. Don't create internal acronyms that may be confused with external acronyms (e.g., CAPA).
- Avoid jargon, slang, and sports metaphors.
- Define the unfamiliar—italicize the first occurrence of unfamiliar terms.
- Place the main information in the main clause.
- Don't overuse pronouns. Specify antecedents.
- Remember that writing is a process and you will progress from drafts to revisions to final versions.
- Use spell checker.
- Use Word readability statistics for summary information about your document. Aim for 10th grade reading level (Flesh-Kincade Grade Level). This can be surprisingly difficult to achieve if you use engineering or regulatory terminology.

Your SOPs are part of a system and may be categorized by process type or function (product development, manufacturing, etc.). Categorization facilitates organization, review, and approval of SOPs. Use the process diagram in your quality manual to create rational categories and a naming structure for your procedures.

All reviewers of documents should look for content, context, format, and grammar:

- Are all regulatory requirements defined and included?
- Are there adequate details on how to do activities?
- Are instructions clear, concise, and compete?
- Is the procedure technically sound and adequate?
- Is the procedure readable?
- Are mistake proofing concepts included?

WARNING

Before making changes to a procedure, take a few minutes to review the change history. Often proposed changes undo a corrective action that was previously put in place. Be very cautious about undoing an improvement or corrective action that was put in place for a good reason.

In conclusion, the QMS ensures that regulatory requirements and quality objectives are translated into processes. The quality system is a network of processes defined in written procedures. These procedures are the very means of communicating to your organization how to perform processes. Without clear, well-written procedures it is impossible to get consistent and predictable outputs. A hierarchy of documents provides structure and alignment of the network of processes. The organizational structure with defined roles and responsibilities enables consistent performance. Your procedures and records are the lens with which regulators will view your QMS. From this chapter, you have tools and concepts to help you translate regulatory requirements and quality objectives into an efficient and effective QMS.

An efficient quality management system

II

Waste and inefficiency in the quality management system

A quality management system (QMS) needs to be efficient in addition to effective. QMS efficiency is about doing things well and optimizing use of resources to obtain the desired quality and compliance results. Efficiency is the ability to avoid wasting materials, energy, money, time, and human effort.

Of course, the regulators do not care as much about efficiency as effectiveness, but other stakeholders do. Employees care a lot about efficiency and feel frustrated and demoralized by burdensome, bureaucratic, and often confusing policies, processes, and procedures. And business stakeholders care about the costs of compliance and quality and impact to overall business success.

An effective process need not reduce efficiency. It need not cause delays or consume more resources. Actually, it is an ineffective process that results in reduced efficiency. It is an ineffective process that leads to errors, mistakes, and nonconformities. It is an ineffective process that yields inconsistent outputs, unpredictable results, and unpleasant surprises. Errors and nonconformities must be corrected leading to rework and delay.

- An ineffective manufacturing process creates nonconformities that may result in scrap, rework, reinspection, or backorder. Combined with an ineffective control plan, nonconforming product may reach the customer.
- An ineffective design control process may result in design requirements that are not met, validations that need to be repeated, redesign activities, extra design reviews, and more. Surprises lead to frustration, missed deliverables, and delayed product launches.
- An ineffective risk management process may result in unacceptable risk, unknown failure modes, higher than expected complaint trends, and even a recall.

An effective process, with a focus on prevention, can result in significant improvement in efficiency and more predictable results. Author, Phil Crosby made the point that it is the **absence** of quality (e.g., nonconformities, audit observations, and failures) that are the real drivers of decreased efficiency.

Quality is free. It is not a gift, but it's free. The 'unquality' things are what cost money.

Philip B. Crosby

Medical Device Quality Management Systems. DOI: https://doi.org/10.1016/B978-0-12-814221-9.00005-1

Inefficient QMSs are characterized by:

- Organizational perceptions that quality and compliance is burdensome, overly bureaucratic, and ineffective
- Frustrations with IT systems (such as corrective and preventive action (CAPA)) that are not intuitive and easy to use
- Constant fire-fighting and crisis management
- Resources are spent more on reaction and not on prevention
- There is excessive waste in terms of nonvalue-added activities
- Inefficiency creates more opportunities for error, ultimately reducing effectiveness
- "Unquality" (as described by Crosby) things such as nonconformities, errors, and mistakes result in unpleasant surprises, project delays, scrap, rework, reinspection, corrective and preventive action, recalls, and regulatory enforcement

Inefficiency of the QMS has many causes, including:

- Reentering of data
- Excessive redundancy
- Confusing roles and responsibilities
- Duplication of effort
- Gaps and overlaps in processes
- Unnecessary handoffs
- Multiple levels of approvals and signatures (i.e., rubber stamp)
- Hybrid systems of approval including hard-copy and electronic signatures
- Hybrid systems of document control
- Waste in the form of increased waiting time and delays
- A long feedback cycle resulting in excessive rework and remediation
- Poor failure investigation or root cause analysis.

These inefficiencies cause employee frustration and demotivation. Crosby referred to these as the "hassles." Even more importantly, the hassles sometimes cause employees to resent the hassle and create their own work arounds. This creates a vicious cycle of noncompliance and resulting quality problems. The solution to this nasty cycle requires QMS efficiency improvement, and attention to the stages of competence (covered more in Chapter 7: Quality is not an Organization).

Duplication of effort is a hassle. Duplication of effort is usually a result of misalignment between process owners and stake holders. Duplicate information is inefficient. More importantly, it also creates a real potential for mismatched information which is a nonconformity and example of ineffectiveness. Double and triple checking are usually examples of duplication of effort and excessive redundancy from crisis mode corrective actions. Corrective actions made in haste often lack thorough root cause analysis. "Band-Aid™" corrective actions are put in place that do not address the root cause and add layers of double checking and duplication of effort. And the real problem is unresolved. This is not only inefficient but ineffective too!

Value-added work is an important concept. In six sigma and lean thinking, value-added work is defined from the viewpoint of the end customer (or regulator), as activities that add value to the product and that they are willing to pay for. In a medical device company, we can also see compliance with regulations as value-added work. By contrast, nonvalue added is considered waste and is work that does not add value to a product, process, or regulatory compliance. Redundancy, duplication of effort, scrap, rework, correction, and corrective action would all be considered nonvalue-added.

CASE STUDY

A VP of R&D wanted his new product development team to keep design control activities separate from business stage gate activities. This caused duplication of effort in planning. The project team leader had to create, update, and maintain two sets of project plans and Gantt charts. Team members were confused about which plan had priority and how activities were linked to each other. Two sets of books created an inability to see the project and activities as a whole. This led to missed deliverables, rework, repeated design review meetings, etc. and required additional updates to both of the Gantt charts. It was a vicious cycle of inefficiency and ineffectiveness.

Some would even say that inspection and audit are unnecessary and nonvalue added from the perspective of the end-customer. Value-added work is considered work that adds value from the perspective of the customer. And the customer perspective says is that if things are done right the first time, there is no need for or value to inspection and audit. Of course, this strict lean concept does not work well in a highly regulated industry where lives depend on product quality, and these activities are required by regulation. But, it should make organizations think about minimizing these activities and focusing on doing things right the first time! Interestingly, ISO 13485:2016 clause 0.3 describes considering processes in terms of added value.

It is well known that 100% inspection is not 100% effective. Reliance on inspection alone will not result in acceptable quality. And it is also known that audit is just a sample and a snapshot in time and is not 100% effective. Reliance on audit alone will not result in acceptable compliance. Prevention of problems is the key.

An important concept here is the feedback loop. Processes may be open loop or closed loop. The difference is the use of feedback to monitor the performance of the process. Closed loop processes have built-in feedback mechanisms. The feedback loop must have enough sensitivity, be timely enough, and continuously monitored so that the need for correction is recognized. Open loop processes do not have the self-checks and real-time monitoring. Open loop processes require independent inspection or auditing to determine conformity to specifications. When compliance issues occur in open loop processes, it may be months or years until these nonconformities are noticed during a Food and Drug Administration

(FDA) inspection. Quality issues may not be known until a call is received from a customer with a complaint. All this time, the same error or problem has been recurring with more and more nonconformities being created.

The feedback loop is the time from when an event, error, or nonconformity occurs, it is discovered, and it is documented as a nonconformity or audit observation. Based on this, feedback is provided, and the nonconformity is dispositioned. The longer the cycle time, the higher the volume or level of nonconformities that occur, and the more mitigation, rework, and remediation that is necessary. When the feedback is provided by an independent step such as internal audit or, even worse, an external regulatory observation, the correction and corrective action required becomes more formalized and rigorous. Minimizing the length of the feedback cycle is essential to reducing the volume of nonconformities and the resulting level of correction and corrective action. Every company needs to think about the length of the feedback cycle in their acceptance activities, inspection points, and audit strategies. See Fig. 5.1.

CASE STUDY

A medical device company received a Form 483 and later a warning letter that included an observation that design validation was not performed under defined operating conditions on initial production lots, units, or batches. The Form 483 observation was made 4 years after the validation was done. During those 4 years, many more validations had occurred. In response to the Form 483 observation, the company had to review all previous validations to see if they had the same nonconformity. For each additional instance they had to assess the risk, determine if revalidation was required, and revalidate.

If this issue had been identified by internal audit only a year or so after the unacceptable audit, the volume of risk assessment and revalidation would have been significantly less. If the nonconformity had been identified via a real-time

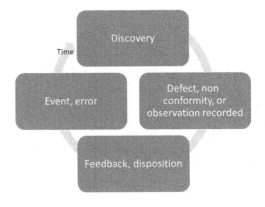

FIGURE 5.1

Feedback loop.

process check, there would have been only been a learning opportunity and no correction or corrective action required. The shorter the feedback loop, the better.

Nonconformities, errors, and drifts in process performance may occur. But, keeping the feedback loop short can minimize their impact. Using errors as learning opportunities to facilitate individual and organizational growth can be considered preventive. But, this must be done as soon as possible after the errors occur. Therefore, strive to keep the feedback loop short and share learning as issues occur. Ways to do this include:

- Mistake-proofing and poka-yoke concepts to minimize error
- Self-check or self-identify systems have the shortest feedback loop
- Successive-check systems where the next step reviews the previous step
- Real-time process monitoring.

Excessive layers of approval (or inspection) do NOT usually increase effectiveness. Indeed, they can actually reduce effectiveness. When an approver sees that many others have already approved (or will approve) a document, they tend to shirk their own responsibilities and approve just because others already have. They assume that the document is acceptable because others have already approved it. But, other approvers are making the same incorrect assumption. This is a huge problem resulting in poor quality work and unnecessary risk. And of course, these additional approvals take more time too. Excessive layers of approval do result in increased cycle time and reduced efficiency.

CASE STUDY

While I was still at GE in 1992, GE pleaded guilty to criminal fraud charges and agreed to pay $69 million in criminal and civility penalties in a massive fraud and bribery scandal involving the sale of nonexistent military jet engines (another highly regulated industry) to Israel. Specifically, key GE employees, together with an Israeli General, falsified documents intended for the purchase, maintenance, and support of F110 engines. The Israeli General pleaded guilty in Israeli court to bribery, accepting kickbacks, and fraud.

I remember being called into a large company meeting to be informed about these issues. What struck me was that all of the falsified documents had been approved by multiple layers of management! There were many, many purchase orders, billing statements, and invoices for engines that never existed and services that were never performed. I remember seeing examples of documents with as many as eight levels of signatures! These examples were shown to employees to demonstrate the responsibilities of approving documents and the consequences of doing it carelessly. It was an eye-opening moment for me and many others.

Approvers of documents, procedures, and records have real responsibilities to ensure the quality, completeness, accuracy, and integrity of anything they approve. They cannot shirk or delegate this responsibility. It is theirs alone. More layers of approval do NOT increase effectiveness. They only give a false sense of security. Use the concept of a RACI (responsible, accountable, consulted, informed) model (see Chapter 15: Alphabet Soup) to guide decisions about who needs to approve documents. Make sure that every approval required has a purpose and clear accountability.

Another source of QMS inefficiency or ineffectiveness is poor use of automation. Be watchful for systems that are so complex that they are not intuitive. Excessive IT systems and automation that is not intuitive can lead to both ineffectiveness and inefficiency. Some CAPA systems are so difficult to maneuver, that additional resources, such as CAPA coordinators, are needed to facilitate entry of information for infrequent users.

Additionally, some companies try to automate processes that are inefficient to begin with. Make sure the processes that you are trying to automate are streamlined, efficient, and effective to begin with. Use lean concepts or value stream mapping to carefully analyze current state and map future state processes. Automation will not solve the problem if your processes are inefficient or ineffective to begin with.

CASE STUDY

A VP of quality refused to use the IT system to approve audit response plans and CAPAs. When her approval was necessary, her staff were required to print out documents for approval. Sometimes it would take days to get her feedback. Sometimes input or comments were made on hard copy, requiring delegation to a staff member to input these comments into the IT system, and then reject the response plan. Approvals were made on paper, and then documents had to be scanned, and attached to electronic records. This bottleneck created additional nonvalue-added work, delays, and additional opportunities for error.

Management that refuses to use IT-enabled systems to review or approve documents sends mixed messages to the organization. They create significant amounts of extra work and opportunities for error. They require personnel to print out documents, deliver them to said management, and leave them for review and approval. Personnel must remember to retrieve them later, scan them, and then manually add the documents into the system. This can also create discrepancies in revisions, timelines, and other approvals. Management needs to walk the talk and use the systems as intended. If IT systems are not intuitive, then fix the IT systems rather than creating additional layers of bureaucracy and inefficiency.

Inadequate processes and procedures are often the cause of problems. Poor process instructions, inadequate level of detail, and unclear expectations leads to errors and nonconformities. And errors require correction, rework, CAPA, and more inefficiency. Unclear roles and responsibilities are a source of confusion, disagreement, and inconsistent outputs. Process problems include poorly defined start and stop points and lack of connection to input and output processes. Use the Supplier, Inputs, Process, Outputs, Customer (SIPOC) approach (see Chapter 15: Alphabet Soup) to clearly define the input and output processes. Work with the owners of the input and output processes to ensure alignment of expectations, inputs, and outputs.

For example, in some companies there is a handoff from a call center to the complaint handling unit to the MDR (medical device report) reporting process. The call center receives initial information, then passes that to a complaint handling unit. The complaint handling unit processes complaints, interacts with other functions to do a failure investigation, and then hands reportable complaints over to regulatory or another group for reporting to regulatory agencies. For all of these steps, how is information handed over? How is subsequent additional information managed? For those complaints that are reportable to the FDA, when does the handoff occur? Is it even necessary? How is it done? What is the signal? What data or documentation is required? Every time there is a handoff from one organization to another, there is a potential for error.

If two connecting processes are not well connected or aligned, there may be a resulting gap (a requirement that was not completed) or overlap (an effort that was duplicated with potentially inconsistent outputs). The gap is a nonconformity (ineffectiveness). The duplication of effort is waste (inefficiency) and the inconsistent output is a nonconformity (ineffectiveness). A swim lane diagram or value stream map (see Chapter 15: Alphabet Soup) are useful tools to clarify steps and define handoffs, with required expectations. Every process owner must make sure that they are aligned with the owners of the input and output processes. Lack of clarity will result in inconsistent process outputs.

Another handoff example is the transfer of a new product design to manufacturing. Design transfer is an explicit requirement of the quality system regulation (21 CFR 820.30(h)). Despite this being an explicit regulatory requirement, many companies still struggle to do this well. Rather than have a clear, standardized process, it is left to individual experience and tribal knowledge, sometimes with unpleasant results. Lack of clarity results in confusion, frustration, and delay. Projects may be delayed, or important activities are missed. To improve efficiency and effectiveness, develop a standard procedure for design and development planning to drive alignment, use standard design transfer checklists, and RACI models or Gantt charts to define required activities with clear roles and responsibilities. An ineffective process will never give predictable results.

Consider mistake-proofing concepts or poka-yoke (see Chapter 15: Alphabet Soup). The intent of poka-yoke is to prevent problems from occurring. Although these concepts are traditionally applied to manufacturing operations, they can apply to transactional processes too.

Mistake-proofing concepts for all processes include:

- Ensure well-defined processes with clear start and stop points.
- Ensure alignment and clear expectations with suppliers (internal and external) of inputs and customers (internal and external) of outputs. Use SIPOCs, swim lane diagrams, or flowcharts to provide clarity for handoffs or connections.
- Use failure modes and effects analysis (FMEA) (see Chapter 15: Alphabet Soup) to predict and prevent problems.
- Ensure clear, well-written procedures with adequate level of detail to enable consistent and predictable results.
- Use checklists or forms to ensure completion of all activities, collect required information, and more easily detect missed steps or information.
- Use automation to further mistake proof processes.

Lack of collaboration is another common source of inefficiency. When different parts of the organization are off working on disparate activities or other priorities, it is hard to improve, or even maintain, quality system performance. Resources are not focused, improvements are not planned and properly managed, progress is not made, and results are minimal. Organizations suffer with inefficiency and often resort to fire-fighting mode, dealing only with one crisis after another.

There is sometimes a vicious cycle between inefficiency and ineffectiveness. Lack of effectiveness is often dealt with by adding additional checks, additional inspections, reviews, and more approvals. Of course, adding these additional steps results in duplication of effort, confusion about roles and responsibilities, and also creates more opportunities for error. This results in more nonconformities. The nonconformities are dealt with by adding more checks and so the vicious cycle continues. This cycle is very apparent when companies are in a crisis situation, resulting in significant changes but unaligned activity.

Overly reactionary QMSs are a huge source of inefficiency. The organization's resources are always focused on the crisis of the day. Firefighting becomes the norm. People get taken off one important project to go deal with the latest fire. The old project stalls and is only partially or never completed. Controls are not put into place to sustain the improvements. And the problem recurs and becomes the next fire, starting the cycle all over again. Start and stop, firefighting, and start again, and do-over modes are all signs of inefficiency that lead to QMS ineffectiveness. Prevention is the key to reducing this source of inefficiency.

Lack of a risk-based approach to CAPA creates a highly inefficient CAPA system. And an inefficient CAPA process means that important things do not get taken care of in a timely manner. The feedback cycle is lengthened. There is more rework and remediation to be done. Resources are scarce and focused on correction and not prevention. That leads to an ineffective CAPA process. There are companies that put everything into their CAPA system. In some cases, CAPAs duplicate activities that are already managed in nonconformance or complaint systems. Everything enters the CAPA systems and stalls. This is a significant problem for companies that are in crisis mode although it can happen anywhere.

Due to lack of prioritization, important CAPAs languish for months, or even years. When brought to management's attention and prodded sufficiently, these aged CAPAs are completed in a rushed and ineffective manner. An already overloaded CAPA system must now deal with another ineffective CAPA, resulting in more rework and delay.

Proper risk management and prioritization is important to both efficiency and effectiveness of your CAPA process. Prioritize CAPAs based on:

- Severity and occurrence of the failure
- Impact to product quality
- Impact to compliance

Another example of inefficiency is separation of CAPA by data source. For example, CAPAs originating from audits are maintained in a separate system with separate rules. I have seen examples of companies that go to great lengths to create a dual IT system with separate features, process steps, and system access. This creates a huge amount of inefficiency with no real benefit. And, it can be actually harmful as companies have more difficulty managing risk and prioritizing improvement activities when they have multiple systems. "Quality audit reports" are explicitly mentioned as a source of data in 21 CFR 820.100(1). Further the FDA explicitly expects medical device manufacturers to take corrective and preventive action to address nonconformities relating to product, process, and quality system. Instead of wasting precious resources trying to hide information about audits, put those resources to work taking corrective actions to known audit issues. Remember, you do not have to show the FDA your entire audit reports, but you do have to records that you are taking corrective and preventive action!

Failure of management to plan for the future and to foresee problems has brought about waste of manpower, or materials, of machine-time, all of which raise the manufacturer's cost and the price the purchaser must pay.

W. Edwards Deming

AUTOMATED SOLUTIONS TO QUALITY MANAGEMENT SYSTEMS

Many medical device companies still use paper based, manual systems for managing documents and records. Paper-based systems are always a challenge in terms of cataloging, storing, protecting, and maintaining documents for long periods of time. On-site storage becomes a problem and then companies must resort to off-site storage. It becomes very difficult to manage, find, and use documents, especially during regulatory inspections. Version control, difficulty finding obsolete documents, documentation errors, and missing documents are frequently seen problems with paper-based systems. There are additional costs to using paper-based systems including increased review and approval time and effort. In short, a paper-based system is not only ineffective but also inefficient.

Companies should carefully consider mechanisms to control and maintain training records. IT systems can really improve efficiency and drive effectiveness too. They can make sure that people are trained before they complete critical processes, they can send out reminders and track compliance. They can be tied to document control activities and ensure that updates and reminders are sent out.

Although the fundamentals of QMS automation have been in place for years, IT solutions to improve efficiency have changed significantly in the recent years. Medical device manufacturers have moved from spreadsheets and paper documents to automated solutions, either home-grown or quality modules in

Enterprise Resource Planning (ERP) systems. Many companies now have disjointed, unconnected systems that do not work well together. This gap is slowly being filled with integrated electronic quality management systems (EQMS). These are essentially a quality management platform with integrated IT architecture and data management solutions to facilitate communication and collaboration. Some of the capabilities being offered include:

- Product lifecycle document control
- Change control
- Supplier quality management
- Complaints management
- MDRs management
- CAPA management
- Product quality planning
- Nonconforming material management
- Audit management
- Employee training
- Acceptance activities
- Automated device history (batch) records
- Real-time management of quality metrics and dashboards
- Calibration and equipment management

Often these capabilities are connected, enabling more efficient quality management and communication with standardized processes and real-time data and metrics. Some EDMS systems also allow collaborative review, editing, and approval of documentation. For example, all members of a product design team can collaborate electronically on a design document enhancing communication and speeding up review and approval.

Many medical device manufacturers choose to automate their quality management systems. Benefits to automated QMSs include improved efficiency, agility, and intelligence (ability to connect the dots).

WARNING

If you do choose to use electronic systems, make sure you use them consistently. Do NOT make exceptions for certain individuals (in particular, upper level management) who do not want to use the IT systems and ask that documents are printed out for them. Then, the documents must be signed manually, scanned, and attached to IT records. Emails are required to document notification times and escalation of issues. This creates new levels of inefficiency and risk. Each time this happens, it creates not only inefficient use of precious resources but significant opportunities for error. If management is uncomfortable using IT systems, then address the root cause and either train management to use the system or make the IT systems more intuitive to use. Do NOT create hybrid systems that are part manual, part electronic with multiple exceptions and exclusions. It just does not work. Additionally, it sets a very poor example from upper level management. This is one place where management needs to walk the talk and use the IT systems as intended.

When implementing technology, have the end in mind. Consider what information and what format you want the data in. How will it be used? What reports will you need? Who are the receivers of the information? How will you make sure that they receive the full benefit of the information? Without an end in mind, you will just be feeding the beast and not extracting useful information. A successful implementation team requires system users in addition to technologists and project management.

Many companies wait too long to switch from paper-based to electronic data management systems due to the investment cost. And they underestimate the long-term savings with switching early on. It can be very beneficial to switch as early as possible to maximize the improvements in efficiency and effectiveness. Additionally, migration of documents and data to new systems becomes more and more difficult the longer you wait to implement one.

WARNING

Although EQMS systems can be very useful when fully installed, many companies struggle to implement them well. Changing to a new IT system can mean significant changes to existing processes. Many companies try to customize the IT solutions rather than using them as is and making their processes adapt. This can create significant difficulties in implementation and validation.

TIP

Medical device manufacturers should implement EQMS solutions as early as possible to minimize implementation/migration costs and reap benefits as soon as possible. Companies that grow through mergers and acquisitions should consider migration to the standardized system as part of their routine integration strategy.

In other cases, the organization is resistant to change. They are slow to adopt IT systems for improvement. They fight automation tooth and nail and stay in a state of happy inefficiency. Change management and stakeholder alignment is needed to bring the nay-sayers along.

CASE STUDY

At one client, the Vice President of the Quality Organization was resistant to using an available electronic system for managing CAPA. He forced his organization to do all CAPA activities on paper. When the CAPA activities were complete, he then had a staff member scan all of the documents and put them into the CAPA system in bulk. Of course, all transactional fields showed the date of entry not the actual date of activities. All entries showed the staff member as the person executing the transaction instead of the actual responder. There were none of the usual ticklers and reminders. The CAPA process was crippled by this ridiculous behavior. Don't be a dinosaur in a world using enabled automated electronic systems!

CONCLUSION

Inefficiency and ineffectiveness exist in a vicious cycle, one resulting in more of the other. Both must be addressed in order to establish and maintain a suitable and effective QMS. A QMS that is resource constrained cannot afford to waste precious resources on inefficiency. The way to break the cycle is to focus on prevention of issues in the first place. Inefficiency is also caused by poorly written procedures, poor process alignment, and nonvalue-added activities.

In today's business environment, medical device companies cannot ignore inefficiency. Key steps to reducing QMS inefficiency include:

- Focus on prevention. Do it right the first time.
- Eliminate nonvalue-added work in the form of excessive handoffs, excessive approvals, reentering of data, and duplication of effort.
- Shorten the feedback cycle wherever possible.
- Improve processes first to ensure they are capable, consistent, and predictable.
- Incorporate mistake-proofing concepts into all processes.
- Introduce automation carefully after processes are improved.

Avoiding common mistakes

6

There are many pitfalls on the road to establishing a quality management system (QMS). Be alert for some of the common mistakes that companies make with their QMSs.

One of the most frequently made mistakes in regulated medical device companies is not having a process to effectively manage regulatory inspections. Depending on where you design, manufacture, and sell your products, you will receive various regulatory inspections. In order to sell medical devices in the United States, you must register your facilities with the Food and Drug Administration (FDA). And this puts you on the list for potential inspections. Being ready for an FDA inspection requires ongoing effort and compliance with your QMS. But, it also highly important to have a plan in place long before the inevitable inspection. See Chapter 16, FDA Inspection Readiness for more detail.

Certification to ISO 13485 is important to marketing products outside of the United States. And it is a big accomplishment to get certified. However, do not make the mistake of thinking that this prepares you for an FDA inspection. An FDA inspection involves another set of depth and rigor over a notified body audit. See Chapter 16, FDA Inspection Readiness to create an inspection preparedness plan.

Mergers and acquisitions are important means of growth for many companies. Do not underestimate the challenge of fully integrating the acquisition. Although companies do due diligence, they are often surprised by later problems. Due diligence is only a small snapshot of the risk picture. Companies are often surprised by much bigger issues over time. My experience is that small companies can sometimes get away with a less effective QMS than a larger company might. I have seen too many cases where, very soon after acquisition, there was a serious quality issue like a recall or a major compliance issue such as a warning letter.

Avoid training gaps in the organization. Culture and training play a huge role in execution of and compliance to procedures. In many companies, training is focused at lower level positions or manufacturing operators. Do not forget to train process owners and functional management all the way to the top. Awareness of regulation history, understanding of metrics, and risk-based decision making are essential skills. Training should be a vital part of improvement and prevention of quality problems, regardless of level, department, or role.

Beware of automation. Automation, done well, is a good thing. In an effort to streamline activities and improve efficiency, many organizations attempt to

Medical Device Quality Management Systems. DOI: https://doi.org/10.1016/B978-0-12-814221-9.00006-3

automate processes like complaint management, corrective and preventive action (CAPA), manufacturing batch records, document control, etc. Automation can bring big gains in productivity and process consistency. But, it is a common mistake to automate processes before they are ready. If the core process has underlying problems, automation will make it worse. Analyze and map your processes first. Identify improvements, stabilize, and then automate. Be cautious about customizing IT solutions or electronic data management systems excessively. Pick a company with a good product and do not attempt to customize excessively. This creates more work, more opportunities for error, and more difficulties in long-term maintenance and control.

Take care when estimating your own capabilities and maturity. It is easy to overestimate your capabilities. Overestimating capabilities will give you a false sense of security. It is better to create clear criteria for your estimation. Support your estimation with data. Use a team-based approach to evaluations. Repeat the review annually as you create the next year's goals and objectives. Self-awareness is a prerequisite to planning.

Be especially careful to evaluate your internal audit capability. Use an independent, well-qualified, external third party to provide a clear evaluation. And no, that external third party should NOT be the FDA! Be proactive in understanding your skill sets. I have seen FDA Form 483 observations for internal audit deficiencies. Usually, these are basic issues such as not doing internal audit or not having a defined schedule. But, I do recall a particularly memorable warning letter in which there were several serious deficiencies for a specific design history file. The next item on the warning letter stated that the same file had just been reviewed in internal audit, which had missed all of these serious issues.

Pay attention to the process approach and process ownership. All the elements of the QMS need to work together in an aligned and harmonious manner. Lack of alignment, gaps, and overlaps in processes can create tears in the fabric of the QMS. Be watchful for signs that functional leaders do not understand or accept their responsibilities for process ownership, control, and monitoring.

Lack of process ownership is one of the main causes of an ineffective and inefficient QMS. When process owners fail to understand and embrace the responsibilities of their role, problems occur. The process is not effectively translated into clear procedures. Personnel do not follow procedure consistently. Output varies. Nonconformities occur. An ineffective and inefficient process is never predictable. And ultimately that hurts business results.

A very common and serious mistake, resulting in an inefficient and ineffective quality management, is failure to use the CAPA system properly. Many companies treat CAPA as a regulatory requirement rather than what it is intended to be—a robust, methodical, closed-loop process to fix problems permanently. They treat CAPA as a burdensome process that is ineffective and hard to understand. They use an IT tool that is not intuitive or easy to use. They cram everything into an overloaded bureaucratic system where everything stalls. There is lack of prioritization and governance. This is known as death by CAPA. Your CAPA program needs a CAPA. See Chapter 15, Alphabet Soup for detail on creating an efficient and effective CAPA process.

Another mistake is not managing CAPA timeframes in a rigorous manner. Some companies have CAPAs that are several years old. It is possible that some CAPAs make take some time to fix properly. In particular, product design changes may take some time. But, aged CAPAs should be the exception and not the rule. Actions should be commensurate with risk! Additionally, all CAPAs that are closed as ineffective deserve management scrutiny. Very old CAPAs, and evidence of closing and replacing an old CAPA with a new CAPA, should be viewed with much suspicion.

Poor root cause analysis (RCA) is another common mistake with CAPA. Sometimes RCA is limited to an obvious symptom. In that case a a quick fix only is applied. The problem may seem to go away for awhile but eventually resurfaces. Or the root cause is incomplete resulting incomplete correction and corrective action. Contributing factors remain unaddressed and result in another problem somewhere else.

When doing RCA for CAPA do not make the mistake of thinking there in only one true root cause. There may be several layers of causes as well as contributing causes. For every recall, there are at least two main causes. The first is the physical cause of the defect, and the second is the failure to detect it before it escaped the quality plan and got out to the customer. Dig deeper. Ask why five times, or more, if necessary. Why was the defect not detected? An inadequate test method? Perhaps a sampling plan that is not statistically valid. Think beyond the physical cause of the failure, and also to contributing causes including poor decision making and human error. Understand if MEDICS immaturity is a contributing cause (see Chapter 8 for detail on MEDICS).

Taking corrective action based on incorrect or incomplete RCA is just a waste of time and energy. It wastes already precious resources exacerbating inefficiency and ineffectiveness. For serious problems, use the rigor of your CAPA process to put permanent fixes in place! Because CAPA is such an important foundational process, an entire chapter is devoted to it. Please see Chapter 15, Alphabet Soup for details on the CAPA process. That chapter includes detail on RCA concepts and a variety of tools that are basics for knowledgeable management with executive responsibility, management representatives, and process owners.

Avoid holding management review meetings too infrequently. Management needs timely and regular information to make informed decisions and prevent trends from becoming full-blown issues. Hold management review with sufficient frequency to assess real-time information, so that actions can be preventive instead of reactive. Management review only once or twice a year is not optimal. Conduct management review in a complete and rigorous manner with clear information and sufficient detail to really understand the health and suitability of the QMS.

The biggest mistake of all is to ignore the warning signs of problems. Every serious Form 483, warning letter, consent decree, or significant quality problem that I have consulted on presented warning signs ahead of time. Yet, management failed to heed the warnings. They ignored evidence and/or took unacceptable risks. They failed to provide enough resources for a suitable and effective QMS. Do not make that mistake. Pay attention to the MEDICS, the capabilities to avoid these common mistakes. They are discussed in detail in Chapter 8, Capabilities and MEDICS for an Effective QMS. Levels of maturity for the MEDICS capabilities are defined, so each medical device can assess their current state.

In conclusion, these mistakes can be avoided. Death by CAPA does not have to happen. CAPA can be the rigorous, methodical process necessary for solving the company's toughest quality problems. Use it when necessary and then use it well. Assess the MEDICS levels to avoid common pitfalls. Pay attention to the details, DO sweat the small stuff, and focus on prevention to create an effective and efficient QMS.

Roles, responsibilities, capabilities

Quality is not an organization

The quality and compliance organization has a bad reputation in many medical device companies. Quality is sometimes seen as the organization solely responsible for ensuring quality and compliance. They are the police force who are out to catch others. Or, they are seen as an organization that is risk-averse, overly bureaucratic, and out of touch with business needs and challenges. An effective QMS requires careful and deliberate actions to prevent these misconceptions.

As discussed in Chapter 3, Establish and Maintain, every organization must determine its definition of and commitment to quality and compliance. Quality and compliance requires a network of processes/functions working toward a common vision of quality and compliance. This requires collaboration and cooperation throughout the company. Quality and compliance cannot be seen as the responsibility of the quality function alone!

The organizations that are most successful in terms of quality are the ones where everyone plays a role and accepts responsibility for quality and compliance. The quality organization plays a guiding role with high-level goals and development of an effective quality system. The quality organization becomes an integrated part of the organization. The quality organization helps to pull together the right team and organizational structure to enable an effective and efficient quality management system (QMS).

The quality organization should guide the organization in developing a QMS, including:

- Determining the regulatory requirements
- Monitoring changing regulatory expectations
- Interpreting and translating regulatory requirements
- Defining the overall QMS structure
- Creating enabling processes such as CAPA, document controls, internal audit, and management review
- Ensuring that regulations are properly identified and translated into processes, documented in procedures, and managed by process owners using metrics and controls
- Identifying threats and opportunities, evaluating risk, and determining strategies for improvement.

Medical Device Quality Management Systems. DOI: https://doi.org/10.1016/B978-0-12-814221-9.00007-5

FIGURE 7.1

Stages of quality and compliance competence.

One of the ways that the quality function can guide the organization is to understand the stages of competence (Fig. 7.1).

This model was created in the 1970s by Noel Burch working with psychologist, Dr. Thomas Gordon. It provides a useful model of learning. It suggests that as employees begin in ignorance, then recognize their ignorance, they consciously acquire a skill, and then consciously use it. Eventually, the skill can be used unconsciously, without even thinking about it. The individual has progressed to unconscious competence. The four phases applied to a context of quality and compliance are:

Unconscious incompetence: Employees are initially unaware of how little they know. They are ignorant of the intent and history of the regulations or how internal policies and procedures must "establish and maintain" regulatory requirements. They perceive quality policies and procedures as restrictive and burdensome. They need training and exposure to strengthen their understanding of and commitment to quality and compliance. They will be heavily influenced by the culture of quality and the norms that surround them. They require exposure to the company vision. The need to hear clear leadership values and quality objectives.

Conscious incompetence: Although an individual does not know how to do something, they recognize their own skill gap. They understand the importance of quality and compliance and actively strive to improve their skills and compliance. But, they do not have the necessary skills and experience to do so. They require training to increase their understanding of how to perform procedures and comply with requirements. They require clear, complete, accurate procedures to follow.

Conscious competence: Individuals are aware of the importance and relevance of their actions to quality and compliance results. They know how to do something but may need to concentrate or break activities into steps. They may experience confusion in their attempts. They may make mistakes.

They need the clarity of good quality policies and procedures to guide their actions. They need reinforcement and guidance.

Unconscious competence: Individuals have had so much practice that everyday compliance tasks are ingrained in their behavior. They can connect the dots. These individuals can communicate and teach their skills to others. They have achieved mastery of the topic. They set an example for others. They provide compelling voices to shape the culture of an organization.

The quality leader needs to recognize that individuals throughout the organization will be at different levels of learning and competence. A quality leader needs to deliberately plan who, what, how, and when to train and educate the rest of the organization through these levels. Organizational structure, development programs, training programs, and stakeholder management are tools that can be used constructively to guide individuals and the organization, as a whole, through the stages of learning. The quality leader needs to provide the awareness, the tools, and the structure for understanding. Train, reinforce, and repeat.

TIP

An effective method for allowing others to learn is to let them observe or participate in regulatory inspections (as a front room scribe, an SME, or backroom support, etc.). Of course, this must be done in a controlled manner (see Chapter 16: FDA Inspection Readiness), but nothing increases awareness and respect for the QMS as much as having to defend it.

It is essential that the quality organization and individuals are independent and able to make appropriate decisions about quality and compliance even if these decisions have a negative impact on business results. It is important to avoid a conflict of interest such incentivizing quality personnel to ship certain products, reduce backorder, or launch new products on time.

The quality organization should also measure and communicate the health of the QMS. They should communicate risks and improvement strategies. And, they may execute specific parts of the QMS such as internal audit which needs to be independent of other process owners. The quality organization creates the overall framework to facilitate quality and compliance throughout the organization. Within that framework, the company can prosper and grow.

World class companies use the concept of a quality council to demonstrate that quality and compliance is everyone's responsibility. The quality council is a senior, cross-functional leadership team promoting awareness and a culture of compliance throughout the organization. This makes it more visible that quality and compliance are the responsibility of all organizational functions. And that leads to better business success.

Someone's sitting in the shade today because someone planted a tree a long time ago.

Warren Buffet

KEY ROLES FOR AN EFFECTIVE AND EFFICIENT QMS
MANAGEMENT WITH EXECUTIVE RESPONSIBILITY

This is one of the most important roles within the QMS. Management with executive responsibility is defined as those senior employees of a manufacturer who have the authority to establish and make changes to the manufacturer's quality policy and quality system. Management is explicitly responsible to establish:

- Quality policy
- Objectives for and commitment to quality
- Adequate organizational structure
  - Appropriate responsibility, authority, and interrelation of all personnel who manage, perform, and assess work affecting quality, and provide the interdependence and authority necessary to perform these tasks
  - Adequate resources, including assignment of trained personnel, for management, performance of work, assessment of activities including internal audits
  - Appointment of a management representative
- Management review to determine the suitability and effectiveness of the QMS
- Quality planning
- Quality system procedures.

It is very important for management to consistently and visibly demonstrate their commitment to quality and compliance. This can be done in a multitude of ways:

- Regularly communicating a focus on the customer
- Respecting and emphasizing compliance with the regulations
- Demonstrating an expectation of continuous improvement
- Starting every employee townhall or communications meeting with a commitment to the customer and providing quality products and services.

> **TIP**
>
> Make sure management with executive responsibility has training records just like every other employee. Training on the quality policy, management responsibility, and management review are critical and need to be completed in a timely manner for this critical role.

Essentially, management has responsibility to establish and maintain an effective QMS with adequate policy, procedures, and resources. Management needs to take *active* steps to review and understand the effectiveness of the QMS. During Food and Drug Administration (FDA) inspections, investigators will attempt to connect violations to individuals that have a duty, power, and responsibility to maintain a suitable and effective QMS. They will consider who has the duty, power, and responsibility to detect, prevent, and correct violations.

There are long standing principles that support this position known as the Doctrine of Strict Liability, Responsible Corporate Officer Doctrine, or simply the Park Doctrine. The Park Doctrine is named after a 1975 United States Supreme Court decision in *United States v. Park* which allowed the government to seek a misdemeanor conviction (and potential subsequent felony) against company officials for alleged violations of the *Food, Drug, and Cosmetic Act*. This applies even if the official was unaware of the violation because the official has responsibilities and authority to prevent or correct the violations. The Supreme Court upheld the conviction of John Park on the theory that executives have an affirmative responsibility to ensure the safety of medical products. The court concluded that the government can criminally prosecute a corporate officer even though that officer did not personally engage in, or even know about, that activity.

The FDA takes the Park Doctrine seriously. The FDA *Regulatory Procedures Manual* covers procedures and considerations for Park Doctrine prosecutions. FDA investigators are trained in the Park Doctrine from early points in their career. They are taught to hold its principles dearly. The FDA places a very high level of importance on management responsibility. The Quality Systems Inspection Technique (QSIT) manual says that a primary purpose of the inspection is to determine whether management with executive responsibility ensures that an adequate and effective quality system has been established. Accordingly, they train their investigators that *each inspection shall begin and end with an evaluation of the management controls subsystem.*

United States v. Dotterwich was another case that went before the Supreme Court. The court upheld strict liability for the president of a drug company. The president was convicted of a public welfare offense for violations of the *Food, Drug, and Cosmetics Act.*

There are other prominent cases holding management responsible. In 2011, four executives of Synthes were sentenced to prison for clinical testing of bone cement without an approved *Investigational Device Exemption*; introducing into interstate commerce a device without FDA clearance or approval; and making false statements to government officials. Federal prosecutors showed that Synthes ignored parts of the FDA approval and labeling process. In 2013, the FDA reached a $1.25 million settlement of a civil money penalty action with Advanced Sterilization Products. Advanced Sterilization Product was required to pay $1.2 million, the President was required to pay $30,000 and the VP of Quality was required to pay $20,000. As I'm writing this book, Alere agreed to pay $33.2 million to resolve allegations that they violated the False Claims Act by knowingly selling unreliable devices.

Warning letters are documented on the FDA website. The FDA also has a database with press releases of compliance, enforcement, and criminal investigations. Individuals, medical practitioners, and medical device companies both small and mighty have experienced the enforcement capabilities of the FDA. These examples are not meant to frighten you (unless you are deliberately trying to circumvent the regulations). They are meant to demonstrate that management has an *affirmative responsibility* to understand and comply with the regulations.

Management can shape culture through both their words and deeds. Individuals in management need to be aware of inconsistencies between their words and actions. If they say they are committed to quality, but act in a contradictory manner, the organization will immediately take notice. Management must walk the talk in order to earn the respect and commitment of the rest of the organization.

Management review is an essential part of making sure that management with executive responsibility has adequate data and information to assess the suitability and effectiveness of the QMS. It must be done at defined intervals and sufficient frequency to ensure they understand issues, status, progress, and the overall health of the QMS.

WARNING

In many companies, management receives filtered, big-picture data that often understates true risks with respect to quality and compliance. Data had been massaged to make it look more acceptable. Dirty little secrets have been left out. And then management is surprised when they receive a serious 483 observation, a warning letter, or must conduct a recall. It is NOT healthy or useful to minimize or hide risks.

Management with executive responsibility needs to be watchful for this. Management, at all levels, can counteract this:

- Make management review a priority.
- Allow adequate time for management review.
- Have a basic understanding of analysis and improvement techniques. Three points in a row going up is not a statistical trend! Read Chapter 15, Alphabet Soup.
- Ask questions and request additional data, if necessary.
- Make data driven decisions.
- Look for indicators of repeat quality issues, repeat or severe audit observations, ineffective CAPAs, profiles of multiple issues, etc.
- Listen for disparate opinions.
- Hold functional managers and process owners accountable for process performance.
- Make sure you have a competent management representative and listen to them! They want to keep you out of jail.
- Understand the information and **heed** the warning signs!

Quality is more important than quantity. One home run is worth two doubles.
Steve Jobs

MANAGEMENT REPRESENTATIVE

The management representative is another critical role. Management with executive responsibility shall appoint a member of management who, irrespective of other responsibilities, shall have established authority over and responsibility for:

1. Ensuring that quality system requirements are effectively established and maintained;
2. Reporting on the performance of the quality system to management with executive responsibility for review.

The appointment of the management representative shall be documented! Make sure that the management representative has a document in their training file that grants them this authority. Further, make sure that the management representative has the appropriate level of authority:

- It should be someone who reports directly to management with executive responsibility.
- They should have the same organizational title and importance as their peers.

TIP

The management representative is a critical role! Expect the FDA to ask for a delegation letter documenting the appointment. They will explore the organizational structure and relationships, asking for organization charts. And they will likely ask for training records. Make sure that these documents are current!

For companies that have multiple sites, there should be a management representative at each site. They should be the senior quality representative at that site. They would be the person that would be responsible to meet with an FDA investigator during an inspection.

Management representatives face many challenges. It can be difficult for quality and compliance leaders and management representatives to deal with business leaders and process owners that lack understanding or respect for the regulations. However, it is possible to educate leaders on the importance of quality and compliance as well as their critical role:

- Create an annual training event, or Quality Day, on quality and compliance.
- Educate them on the concept of affirmative responsibility or the Park Doctrine.
- Have them share their personal experiences, stories, and lessons learned about quality, compliance, and customer satisfaction.
- Share news and stories about other companies' quality problems or compliance issues.
- Share lessons learned from inspection results.
- Keep them abreast of threats, challenges, changes, and opportunities within the industry.

- Create a Quality Council (group of senior cross-functional leaders) to champion and guide quality and compliance.
- Train them on the value proposition for quality and compliance (see Chapter 10: Shifting From Cost of Quality to the Value Proposition).
- Give them a copy of this book!

Prepare management review. Management review shall be conducted at defined intervals and suitable frequency. Management review shall include relevant information about the health of the QMS. Conducting an effective management review that paints a clear picture of the health of the QMS is one of the most vital responsibilities of the management representative. See Chapter 4, QMS Structure for additional detail on preparing management review.

WARNING

Red flags to be watchful for include fear of sharing information with management and attempts to minimize the severity of issues and risks. Also watch out for management that dismisses or makes light of risks and concerns. I have seen all of these behaviors and the destructive consequences.

PROCESS OWNERSHIP

In keeping with the process approach described in Chapter 4, QMS Structure, we must determine roles and responsibilities for process ownership. As each organization creates their process structure and defines the processes in the Quality Manual, they should designate process owners and quality and compliance partners. Every process should have a defined process owner. Most processes need not, indeed should not, be owned by the quality organization. The process owner should be a subject matter expert (SME) and is responsible for:

- Understanding the applicable regulatory requirements
- Translating regulatory requirements into documented procedures
- Providing SME in the topic
- Providing training
- Understanding process inputs and outputs
- Determining connections to other processes and managing those stakeholders
- Identifying and communicating best practices to optimize process effectiveness and efficiency
- Defining process performance metrics
- Monitoring process performance and consistency
- Communicating and sharing process performance metrics with quality partners and in management review
- Continuous improvement and, if necessary, corrective action
- Ensuring records document the effectiveness of the process and can withstand regulatory scrutiny.

The quality organization need not be the owner of all the processes! In fact, it is better if the quality organization is NOT the owner of the processes. The process owner should be the organization or function that is the key stakeholder or executor of the process. For example, Product Development should be the owner of the Design Control process. Operations owns Production and Process Controls. Supplier Management owns purchasing controls. The quality organization itself may own enabling processes such as management review, internal audit, and CAPA. As the organization that must implement and comply with the regulations, the process owners should be the ones that define the process they follow.

The process owner must have sufficient organizational authority to ensure process performance. The process owner's own performance measures and compensation should be tied to process performance. But, remember it is vital to include quality performance in addition to schedule or cost metrics.

CASE STUDY 1

An R&D organization was the designated process owner for Design Controls. The VP of R&D shared his performance goals with the organization. The goals listed the specific products to be launched with clear expectations to be on time and at budget. When challenged to add a goal for quality, he replied that quality was "implied." This sends an inappropriate message to the organization. Why is quality implied when cost and schedule were explicitly emphasized? This organization was plagued with mistakes and surprises which resulted in rework which actually delayed product launches and resulted in cost overruns.

CASE STUDY 2

Another R&D organization created a self-evaluation process to monitor product development quality and compliance real time. They had well-respected experts in the organization review and score documents at designated milestones during the development process. They used any issues found as teaching moments. Immediate correction was taken. Overall, the projects using this approach had fewer mistakes, less rework, more repeatable timelines, and better product launches. Long term, the organization developed well-trained process experts and a more consistent and predictable process.

The process owner must ensure that the four pillars of process ownership are in place:

- **P**rocess and Procedures—define and document adequate procedures. Identify enabling infrastructure or IT technology needed for flexibility, reliability, and consistent performance.
- **P**erformance management—determine metrics and structure for process monitoring and performance.
- **P**eople—establish organizational structure, roles and responsibilities to adequately execute the process, and define and meet training needs.
- **G**overnance—create the governance structure for process management, control, and improvement.

A hall mark of an excellent process is the use of in-process checks to self-identify issues. By self-identifying issues in real time, nonconformities are immediately corrected and fewer nonconformities are created. If problems are not self-identified, they may be found later in internal audit, corporate audit, or even by external regulators and agencies. By then, there are many, many more nonconformities that need to be corrected. Rework is extensive. Issues found through audit require the rigor of a closed-loop CAPA process.

CASE STUDY

A Complaint Handling Unit (CHU) recorded customer calls to ensure information was adequately captured and transcribed into written complaint files. During an FDA inspection, the investigators became aware of these recordings and listened to them intently for hours and hours. Ultimately, they found discrepancies between the recorded calls and the written records. The company received a 483 observation for failing to document all relevant complaint information. In order to address this observation, the company had to go back and review all the recordings (an extensive number) to make sure that the written records were complete and accurate. This caused an enormous amount or rework, utilizing significant resources and time. This could have been easily prevented by regular in-process checks by the process owner (CHU manager or delegate). A weekly sample and review of the recorded calls could have been conducted. A control chart could have been used to document nonconformities and trends. Discrepancies could have been immediately corrected and recurrence prevented. Poor handling of calls could have been used for educational purposes (that was the original intent but never realized). The recordings could have been deleted after this in-process check. It would have been much better if this process had employed some very simple in-process checks.

The process owner does not need to do all of this by themselves. But, they do need to be the process champion. The process owner needs to have organizational authority and/or be higher level management to ensure the process gets appropriate attention and resources. The process owner is the driving force for an effective and efficient process to optimize value for the company. A process owner may employ process managers or process engineers for the more operational aspects of process ownership. They may delegate more routine tasks such as actually drafting standard operating procedures (SOPs). But, they do need to review and approve procedures. They need to walk the talk and ensure proper process performance. They need to be a *visible* champion for the process.

The process owner should also define the governance structure for the process. Without governance, the process becomes less efficient and effective. Improvement initiatives and changes become disconnected. Performance stalls and stakeholders become disillusioned and/or frustrated. Good governance includes identifying the process owner and stakeholder(s), metrics, process monitoring, providing training and education, and identifying and championing any needed improvement activities.

Some companies use the concept of Centers of Excellence (CoE) to define and govern processes. It does not matter what you call it, as long as you effectively govern processes. Process owners/functional leaders can also make use of maturity modeling to define standards of excellence. Using a system of assessment, improvement plans, training, and reward and recognition, they can drive process excellence.

Process owners need to ensure preparedness for FDA or other regulatory inspections. The process owner (or delegate) should prepare an inspection preparedness checklist (see Chapter 16: FDA Inspection Readiness). Use the Quality System Inspection Technique (QSIT) guidance from the FDA to understand expectations and create a comprehensive checklist. Identify personnel at each site that have the skillsets to be an SME during an inspection. The process owner needs to lead any necessary Form 483 responses and improvement efforts.

Lack of process ownership is one of the main causes of an inefficient and ineffective QMS. When process owners fail to understand and embrace their responsibilities, then processes falter. In that case, the quality organization sometimes has to step in and act as a police force. This starts a vicious cycle. It is recommended that medical device companies conduct training on process thinking and ownership in order to optimize process performance throughout the QMS.

Be a yardstick of quality. Some people aren't used to an environment where excellence is expected.

Steve Jobs

INDIVIDUAL ACCOUNTABILITY

Individuals throughout the organization have important responsibilities that they must understand and embrace. These include:

- Understand the quality policy and objectives
- Understand and comply with applicable procedures
- Complete, maintain, and protect records
- Follow good documentation practices (GDPs) and safe writing standards
- Understand complaint reporting requirements (they must immediately report any alleged complaints that they become aware of to the CHU because the clock starts ticking when *anyone* in your organization becomes aware of an alleged complaint)
- Complete required training on time.

Quality and compliance expectations for all individuals need to be integrated into job descriptions, performance metrics, individual goals, and appraisal systems throughout the organization. Quality and compliance needs to be a crystal-clear expectation for every employee.

The Quality System Regulation has clear expectations for personnel and requires that:

- They have the necessary education, background, and experience to ensure all activities are correctly performed
- Training needs are identified and fulfilled
- Personnel are made aware of defects and errors.

ISO 13485:2016 adds clarity by requiring that "personnel are *aware of the relevance and importance of their activities* and how they contribute to the achievement of the quality objectives." Accountability in the workforce needs to be nurtured. Accountability is not about burden, blame, fault, praise, credit, shame, or guilt. It starts with individuals understanding their roles, responsibilities, relevance, and importance combined with a desire, willingness, and training to do the right thing. It is inspired by customer focus and a passion to make sure products help customers, save lives, decrease pain, and protect families.

QUALITY CASE STUDY

When I was a young engineer, the president of the company regularly held town hall meetings which featured customers who shared their real-life stories of how our products had improved the quality of their lives, reduced pain, saved their child or other loved one, or restored their health. These town hall meetings were hugely impactful sessions and employees always left them feeling invigorated and deeply committed to ensuring safety and effectiveness of our products. Hearing these stories about the products that I personally worked on made me feel so proud, so important! And I embraced my responsibilities to make sure the products were safe, effective, and worked properly every single time. In these meetings, executive management emphasized their commitment to developing new products, innovation, quality, and compliance all focused on helping to improve the lives of customers. This one simple, but very powerful, step helped to positively shape the culture of the organization and inspire all individuals.

See it, own it, solve it, do it. Examples of individual accountability, innovation, performance, and quality improvement can and should be recognized and rewarded. Many companies have formed recognition and reward systems for quality improvement. Quality improvement awards for all levels of employees, or even for suppliers or other stakeholders, can be a significant way of promoting and reinforcing individual responsibility, accountability, innovation, and continuous improvement.

Of course, this all requires an effective training process. Training must be considered, defined, and implemented for every individual in the organization. Like everything else in the QMS, records of training must be maintained. Medical device manufacturers must address training needs including:

- Training plans for all job descriptions.
- A new-hire orientation program so that every new employee receives timely training on the basics of the quality policy, objectives, QMS, and individual expectations.
- Periodic refresher training of key quality responsibilities (e.g., quality policy, responsibility to notify the CHU if you become aware of a potential complaint, etc.).
- Retraining as necessary for corrective and preventive action.
- Determining and monitoring the effectiveness of training.
- Certification for critical roles.

> ### CASE STUDY 1
>
> Routinely, when conducting internal audits, auditors ask for training records from individuals. In one case, a particular individual could not provide his training file in a reasonable time frame. He kept hesitating, stalling, and saying he was looking for it. When pressed, he finally produced his training file after 2 days. Approximately 75% of his self-training activities had been completed the day before. This is poor form and something that managers need to watch for. FDA investigators can see through these games too.

> ### CASE STUDY 2
>
> During FDA inspections, certain key individuals such as the management representatives are routinely asked for training files. This should not be a surprise. Yet, I remember a relatively high-level manager panic when the FDA showed up and asked for her file. Don't be the one that is surprised and unprepared! Walk the talk.

When you make changes to high-level positions (e.g., Management with Executive Responsibility or the Management Representative), have a plan for immediate training needs such as the quality policy and management responsibility. These are training needs that cannot wait. Management with executive responsibility needs to be aware immediately of these critical expectations.

It is good practice to review training needs and fulfillment periodically for all company employees. This can easily be tied to annual performance reviews and an updated training plan for the upcoming year. And the training process, just like any other process, should have performance metrics.

Quality means doing the right thing when no one is looking.

Henry Ford

CULTURE

A positive culture of quality and compliance has a huge impact on every person in the company. Culture can be shaped and nurtured. Take deliberate steps to understand and influence the culture of your organization in a positive way.

Culture is influenced by sharing beliefs and values in the form of the quality policy and objectives. The culture includes the language of quality used to communicate. These are translated into plans, projects, and all the way to performance objectives for every individual in the organization.

> ### CASE STUDY
>
> The Johnson & Johnson Company is famous for its credo, written in the mid-1940s by Robert Wood Johnson. The credo places priority on customers first, then employees, then communities,
>
> *(Continued)*

and finally stockholders. See the Johnson & Johnson website for full wording. As a former Johnson and Johnson employee, I remember quite well the importance of the credo. I remember continual management emphasis on the credo and how it guided organizational values and actions.

The credo was tested in 1982 when seven people in Chicago died after taking a Johnson & Johnson product, Tylenol, that had been tampered with. The tampering occurred after the product reached the shelf in the market. The capsules were removed from the shelf, laced with cyanide, and then put back on the store shelf. Although Johnson and Johnson knew that they were not responsible for the deaths, they used the values outlined in the credo to make decisions on how to handle the crisis. Putting customers first, Johnson & Johnson's McNeil Consumer Products subsidiary acted quickly and conducted a recall of an estimated 31 million bottles of Tylenol at a huge financial loss.

Tylenol was reintroduced to the market months later with the first tamper-resistant packaging. Additionally, the Johnson and Johnson subsidiary, McNeil, promoted caplets which are more difficult to tamper with. The killer who deliberately laced the product with cyanide was never found. But, with the safeguards in place, Tylenol regained its market share and became a symbol of trust. The Tylenol crisis soon became a model of crisis management and corporate responsibility.

Management with executive responsibility needs to be aware of the culture. They need to actively shape the organizational culture, values, and expectations for all roles. A culture of quality and compliance, customer focus, continual improvement, and individual accountability can be nurtured by management. Additional information on culture and values can be found in the key capabilities (MEDICS) section in Chapter 8, Capabilities and MEDICS for an Effective QMS.

Quality is the result of a carefully constructed cultural environment. It has to be the fabric of the organization, not part of the fabric.

Phil B. Crosby

In conclusion, an effective and efficient quality management system is the responsibility of everyone in the organization. Management with executive responsibility has affirmative responsibility to establish quality policy and objectives, provide organizational structure and responsibilities, and provide adequate resources. Responsibilities also require appointment of a management representative with the authority and responsibility to establish and maintain a suitable QMS. A suitable QMS also requires process owners with clear accountability for providing SMEs to translate regulatory requirements into documented procedures, with appropriate detail. Process owners must also monitor process performance; enable the process with an appropriate organizational structure and training; and provide governance for process management, control, and improvement. Individual contributors all have a role in understanding quality policy and objectives, complying with procedures and GDP standards, completing required records, and training. These roles are all wrapped together in the fabric of a culture of quality and compliance.

Capabilities and MEDICS for an effective QMS

8

Problems in a quality management system (QMS) create more problems. Ineffectiveness creates inefficiency. Inefficiency creates ineffectiveness. The vicious cycle can be broken with hard work and attention to critical QMS capabilities.

Medical device manufacturers need to have six key capabilities to create an effective and efficient QMS. Without these capabilities they cannot thrive, or even survive, in a competitive and ever-changing industry. Of course, there are many other capabilities that a medical device manufacturer must have (e.g., technical capabilities, manufacturing, etc.) but this book focuses on those capabilities essential for an efficient and effective QMS. These six key capabilities, or Quality System **MEDICS**, essential for a healthy QMS are:

1. **M**onitor—Ability to monitor, measure, and analyze the health of the QMS
2. **E**mbrace—Ability of companies to embrace a culture of quality and compliance
3. **D**efine—Ability to define risks and prioritize issues and actions
4. **I**dentify—Ability to self-identify problems
5. **C**orrective and preventive action (CAPA) and Improvement—Ability to fix problems robustly and permanently
6. **S**hare and Communicate—Ability to share and communicate key information in a transparent manner.

Food and Drug Administration (FDA) inspections focus on regulatory compliance but do not identify manufacturer capabilities to improve quality. However, medical device companies MUST consider their capabilities to comply with the regulations and achieve higher quality outcomes. The MEDICS should be considered key success factors for enabling an effective and efficient QMS. Consider these capabilities as you develop your quality strategy and plans.

First, know thyself...

For each of the MEDICS discussed next, three levels (learning, surviving, and thriving) of maturity are described. As you read them, you may find that your company has maturity characteristics at more than one level. That is normal. Look for general trends and areas for improvement. Each company must decide what level of maturity is necessary for the individual situation, depending on the starting point and company aspirations. Decide what MEDICS maturity is necessary for future business goals and quality objectives. You may decide that your

Medical Device Quality Management Systems. DOI: https://doi.org/10.1016/B978-0-12-814221-9.00008-7

ultimate goal is to be at a thriving level for each of the MEDICS. Or perhaps your immediate goal is just to get to succeeding level or the next level.

You will also find warnings and red flags that indicate serious problems. The red flags present unacceptable risk and could result in unacceptable product quality or regulatory enforcement. Red flags should be clearly articulated and brought to the attention of the management representative.

Each of the MEDICS also has a future trends section that should stimulate some ideas for leading edge capabilities. Perhaps your company will become the leading edge and others will want to benchmark with you. Some companies may wish to have leading edge capabilities to position themselves for growth and competitiveness in the future.

Additionally, the concepts of maturity modeling will be discussed more in Chapter 11, Maturity Modeling in Medical Device Companies. You can use the MEDICS maturity levels to develop more detailed maturity models and improvement plans.

Evaluate your current state and targeted future state. From there, define the needed actions and plans to get to the targeted future state. These may include multi-year plans and should be included in your quality objectives.

Fig. 8.1 provides an example of a self-assessment summary of QMS MEDICS maturity. You can identify your current state and desired future state. It can be helpful to look at MEDICS maturity from this viewpoint. This can summarize a more detailed analysis using the tables later in this chapter. Use the more detailed MEDICS tables below to identify specific characteristics or behaviors that need improvement.

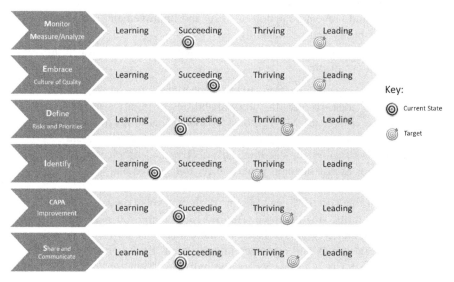

FIGURE 8.1

QMS MEDICS capabilities.

MONITOR

M is for monitor and measure. ISO 13485:2016 clause 8 (Measurement, analysis and improvement) sets clear expectations for this and rightly so. The quality system regulation (QSR), in 21CFR820.100 (a)(1), also sets expectations to analyze processes and other data. Therefore, this is not only a key capability, it is also a regulatory requirement!

Medical device manufacturers need to have the capabilities to monitor the health of the QMS. Process controls, metrics, and analytics are essential tools for an efficient and effective QMS. Metrics are necessary to measure the suitability and effectiveness of the Quality System, inform management through Management Review, monitor product quality, customer satisfaction, regulatory compliance, and the impact on business results. Management has a responsibility to ensure an effective QMS, including adequate resources. Metrics are an essential ingredient for doing this. It takes considerable effort to create a system of good metrics, but the rewards are significant. Data driven organizations make better decisions. In the words of renowned engineer, consultant, statistician, and professor Dr. W. Edwards Deming, "In God we trust, all others bring data." An effective QMS depends on sharing metrics and data in a clear, transparent manner.

Part B of the QSR requires that the management representative shall report to management on the performance of the quality system. And management shall review the suitability and effectiveness of the quality system at defined intervals with sufficient frequency to ensure that the quality system satisfies the requirements of the regulation and the company's own quality policy and objectives. This requires a system of well-designed metrics.

ISO 13485:2016 clause 5.4.1 Planning—Quality Objectives requires that "Top management shall ensure that quality objectives, including those needed to meet applicable regulatory requirements and requirements for product are established at relevant functions and levels within the organizations. The quality objectives shall be *measurable* and consistent with the quality policy." The phrase "relevant functions and levels" makes it clear that a system of metrics is needed to cover all aspects of the QMS.

ISO 9000, other standards, and awards like the Baldrige award also establish expectations for performance metrics and criteria for how an organization selects, gathers, analyzes, reports, and acts on data to improve performance. ISO 13485:2016 puts a strong emphasis on evaluating risk when making quality and compliance decisions. Again, this requires a system of metrics. And lastly, having good metrics just makes good business sense. It makes good business sense to have metrics to make sure your QMS is both effective and efficient.

Obviously, monitoring is a key capability. Companies that cannot do this well will never be able to create an efficient and effective QMS. Therefore, we need to consider the characteristics of this key capability and explore how companies can achieve mastery. We need to define the processes and sources of data to monitor as well as what elements of those sources and how to monitor them.

Companies with beginner level capabilities have few or no metrics, definitions are not clear, and metrics are not linked to quality objectives. They do not have expectations for appropriate statistical techniques. More mature companies have defined metrics but still experience problems with inconsistent use, manual data manipulation, and associated errors, omissions, and delays. Higher levels of maturity include clear metrics linked to quality objectives. A governance board ensures good definition, real-time collection and use, transparency of data, and consistency in making informed decisions. Analytics are used for predictive and prescriptive purposes.

Quality objectives need to be measurable and require good definition. Quality objectives commonly have three performance aspects:

1. Effectiveness measures the ability of a process to achieve its intended result. ISO9000 defines effectiveness as the extent to which planned activities are realized and planned results are achieved.
2. Efficiency measure the resources required for a process to achieve its intended result. ISO9000 defines efficiency as the relationship between the result achieved and the level of resources used.
3. Cycle time measures the duration of a process to meet its intended result.

It is important to maintain a balance between these aspects of metrics. "What gets measured gets done" said Tom Peters in the 1982 best-seller *In Search of Excellence*. There is a lot of truth to this, but it can be a double-edged sword if you focus only on one or two of these aspects. For example, focusing only on cycle time, speed, or schedule can have negative consequences to the quality of the output. Additionally, the Hawthorne effect tells us that just by measuring or observing a process, we can change the behavior of the process. Measurement and metrics must be well-done and balanced to monitor and understand the health of the QMS and to drive the right results. So, it is necessary to set up a system of metrics.

Factors to consider when designing your metric system are:

- Connecting metrics to quality objectives
- Size, structure, and complexity of your organization (i.e., the "relevant functions and layers" described in ISO 13485:2016 5.4.1)
- Mechanisms and ability to gather and analyze data (e.g., automated vs. manual systems)
- Organizational maturity and quality objectives
- Making data available for real-time (or near real-time) use and decision making
- Using data to monitor process performance
- Using data to define risks and prioritize improvement activities
- Transparency of data to executive management.

ISO 13485:2016 clause 8.2.5 Monitoring and measurement of processes, requires that "the organization shall apply suitable methods for monitoring, and as appropriate, measurement of the quality system processes. When planned results are not achieved, correction and corrective action shall be taken." This emphasizes the process approach discussed in Chapter 4, QMS Structure.

Your QMS structure, as documented in your quality manual, should define the key processes to be measured for performance and improvement. Key performance indicators (KPIs) are used to understand the performance of each of the processes within your QMS. Utilize the process structure you have defined in your quality manual when creating your metrics to ensure all processes have performance metrics.

Metrics can be a double-edged sword. When poorly designed and implemented, metrics can lead to problems and undesirable behaviors. It is also important to be aware of errors and omissions as you create your system of metrics. Common problems with metrics can include:

- An overly complicated set of metrics can consume significant resources for reporting purposes.
- Unclear definitions can lead to unclear trends and comparisons (apples and oranges comparisons). It also allows organizations to "massage" the data to make it look more favorable (a big mistake).
- Manual data entry, analysis, and transfer of information can lead to errors and omissions. It also takes a significant amount of time to process data which leads to late information.
- Inconsistent application and use of metrics also leads to mistakes and unclear trends.
- Unbalanced metrics can drive the wrong behaviors. For example, always focusing on cycle time can lead to problems with quality performance.
- Measuring everything just because it can be measured creates too much noise and leads to uncertainty and confusion.
- A cumbersome data collection process can create a long lag time, meaning you are acting on old information.
- Metrics that are inconsistent with overall goals and objectives can result in poor results.
- Goals that are too easy or too hard drive unacceptable results.
- Frequency/cycle of metrics is inappropriate for the current situation.

Metrics can and will evolve over time as your QMS matures. Because organizations usually focus first on what is easy to measure, metrics often go through the 3Es:

- Extent—Example: How many CAPAs are open?
- Efficiency—Example: Are we getting the CAPAs closed? How old are they?
- Effectiveness—Example: How effective are CAPA actions?

Organizations that are immature focus on metrics to determine the extent of something (such as the number of audit observations open). Then they realize this is insufficient to manage the process, so they add additional metrics that are easy to gather (such as percentage of audit observations closed). But, focusing on these things alone can drive the wrong behaviors and results in activities that are not well-done, not compliant, or ineffective. More mature companies also include metrics to measure process performance and quality of the output of processes (such as overall decreased trend in audit observations and number of repeat observations).

A well-designed system of metrics gives thought to the processes within your QMS as well as the customers of your QMS. Of course, the end-customer of your products is critical and also consider other stakeholders in the output of your QMS such as business stakeholders and regulators. A balanced set of metrics reflects all of these customers:

- The end customer means the patients, doctors, hospitals, nurses, and users of the medical devices. End customers are interested in product quality, safety, reliability, and the cost impact of efficacy.
- Regulators responsible for monitoring and enforcing the regulations are also a customer of your QMS. Metrics indicating quality system compliance are valuable in addition to the product quality metrics already described.
- Business stakeholders such as employees, management, and shareholders are also interested in the performance, efficiency, and suitability of your QMS.

Table 8.1 provides examples of metrics in these categories.

As organizations grow and become more complex, they need to be able to combine, roll-up, and escalate metrics in a meaningful manner for different users and layers of management review. This creates challenges in terms of collecting and summarizing data in a timely manner. Consider your organizational hierarchy when designing your metrics process. Metrics for a manufacturing line need to roll up to plant metrics, then business unit metrics, and then company-wide metrics. These metrics need to be appropriate for the level of management reviewing them. Indexes are a good way of rolling up metrics.

Another challenge is creating leading rather than lagging metrics. A tool called a CTQ (Critical to Quality) tree can be used to establish the relationship between lagging and leading metrics. For example, the number if recalls for a company is important and easy to measure. But, waiting until the end of the year to count how many recalls is not very useful in terms of monitoring your product quality. A more leading metric is required. So, use the CTQ tree to define the factors (number of product defects and ability to detect them internally) that result in recalls. Number of defects produced could be evaluated in terms of process capability, scrap, yield, or rolled throughput. The more defects that are created, the more likely that some will escape your detection systems and get into the field. One could also measure number of complaints, volumes, trends, and new failure modes. And your ability to detect defects depends on having good acceptance activities, inspection test methods, calibrated gages, acceptable sampling plans, etc. (See example in Fig. 8.2).

Table 8.1 Metrics Examples

Balanced Scorecard		
Key Process Indicators		
Compliance Metrics	**Product Quality Metrics**	**Business Metrics**
Records demonstrate acceptable performance of activities	Products are safe and effective	Market share
Cycle time of activities (aging or timeliness)	Products are used per defined user needs and intended uses	Customer trust
Activities comply with procedures	Products are distributed within specification	Prevention costs
Procedures are implemented	Products manufactured within specification	Cost of appraisal (see Chapter 10)
Personnel are trained to procedures	Robust and reliable design specifications	Cost of poor quality (see Chapter 10)
Regulatory requirements are completely and correctly translated to procedures	Design outputs meet input requirements	Culture of quality index
Regulatory requirements are identified and understood	User needs and intended uses met	
Examples		
Audit observations (number, severity, repeats, responses on time, actions on time, aging, closed as effective)	Recalls (total number, % product retrieved, timeliness). Failure investigations on-time.	Cost of internal failure
External observations (Form 483 observations, observations per inspection, notified body major observations)	Product holds, near misses, escapes	Cost of external failure (complaints, MDRs, recalls)
Audit remediation percent complete, residual risk	Nonconformances (number, rate, type)	Impact of complaints on good will, sales, market share
CAPAs (total number, total opened in a month, total closed in a month, average and median age, % closed effective, % aged CAPAs)	Complaints (volume, rate, trend, new failure modes, average and median age)	Cost of compliance failure and mitigation (Form 483 responses, warning letter response, consent decree response)
	Medical device reports (number, rates, trends, new failure modes, reporting timeliness)	Training investment
	Process capability, Cpk, Ppk	Cost of prevention
	Yield, rolled throughput yield, scrap, or DPM	Culture of quality results/ index

(Continued)

Table 8.1 Metrics Examples *Continued*

Balanced Scorecard		
Key Process Indicators		
Compliance Metrics	**Product Quality Metrics**	**Business Metrics**
	% Suppliers with quality agreements	Employee engagement index
	% Suppliers with acceptable performance ratings	Customer satisfaction index
	Adequacy of acceptance testing (sample size, GR&R, etc.)	
	In-process controls	
	Batches right the first time	
	Validations right the first time	
	Design transfer acceptability checklists	
	Design reliability index	
	Actual versus predicted failure modes and rates for new products launched.	

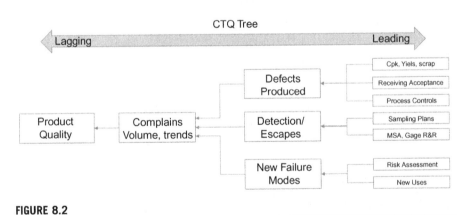

FIGURE 8.2

CTQ tree.

Another useful tool for creating your system of metrics is to use a Fault Tree Analysis (FTA) to determine the relationship between factors. The FTA uses the concept of logic gates (and gates, or gates, nor gates) and diagrams to determine relationships and probabilities. Using these logic diagrams, you can calculate probabilities to determine your risks.

Over the last few years, we have seen the FDA work with industry to create standardized metrics. Ultimately, the FDA will use these metrics to identify risks and prioritize their inspection activities.

WARNING

Just because it is easy to measure something, does not mean it is useful to measure it. Considerations for metrics:

- Is it readily available/easy to track (manual vs. automated systems)?
- Is it objective?
- Is it scalable? How do you average, or escalate it? (e.g. averaging all complaints makes it impossible to see which products are improving or not). Can you segment it later for clarity?
- How do you trend it over time?
- Is it well defined? Does it have a standard interpretation?
- Is it high priority or linked to quality goals and objectives?
- How does it work within the metrics hierarchy?
- Is ownership/reporting clear?
- Can/should the data be normalized?

Consider your processes and sources of data and the elements of the process to measure. You need to consider the 3Es for each data source to ensure you have a balanced scorecard and drive the right behaviors.

WARNING

Be concerned about inconsistent use of metrics or attempts to alter definitions or usage. For example, I've seen organizations with metrics for CAPA actions on time. Some plants or business units report actual data pulled straight out of an automated data base that defines actions as complete when the task owner completes the activity AND it is approved by the quality organization. Other business units "massaged" the data to include actions that were completed but NOT yet approved by the quality organization. They manually changed data extracted from the automated data base to do this. This was done in a deliberate attempt to make the data look more favorable. This unacceptable behavior makes it impossible to compare performance (apples to apples) of different organizations; performance against goals; and trend performance over time. It's also an inefficient use of precious resources to manipulate data that could be more easily extracted with IT business tools. It also increases the frequency of errors and miscounts because the data must be manually adjusted again the next month to reverse the count. Where possible, pull data directly from automated systems to prevent alteration of data! Altering data in an attempt to make it look better is completely unacceptable behavior in a medical device company.

Another bad behavior is when the action owner waits until the last minute to complete his actions. Then the approver(s) from quality or other functions don't have sufficient time to review and approve the actions. Actions are not done until they are *completely* DONE, including acceptable review and approval! Allow appropriate time for review and approval.

Setting up a system of metrics to measure the health of your QMS should include the following steps:

1. Establish a data governance board. This board should include representation from quality/compliance, process owners, and IT. Be sure to include someone with statistical skills such as a statistician, Six Sigma Blackbelt, or a data scientist. Data science is a rapidly developing field. Include someone from IT who can help integrate automated data management and business intelligence tools into your process. Be sure to set a good foundation. This board will be essential in getting stakeholder alignment throughout the organization. The data governance board should also determine requirements for metrics and analysis training as applicable. Determine governance rules for where data is generated, its architecture, and how it is administered including security and access.

2. Conduct an assessment of your current state process. An independent third party can bring a fresh set of eyes with new insights and expertise. Map the current state process including how many layers of data, spreadsheets, analysts, and interpretations there are. Identify sources of data (make sure these are consistent with your sources of data for CAPA). Analyze errors, omissions, inconsistencies, and delays.

3. Map and design your new data management process. Identify top level goals and use the critical to quality (CTQ) tree to get to more leading metrics. Show clear linkage. Define relevant functions, levels, reporting structure, and hierarchy. Define the reporting cycle and link it to management review timelines. Establish owners/stewards of the data. Owners should not necessarily be the quality organization. It is more effective to have the process owners be the stewards of the data for their process. Ensure metrics are well defined with clear goals. Design reporting formats and expectations for well-dressed metrics.

4. Utilize IT/business solutions to manage and prioritize large amounts of data. Although many companies start with simple Microsoft Excel™ spreadsheets, there are tools that can manage large amounts of data more easily and help you to analyze it easily and rigorously. Determine visualization standards. IT data management tools become more and more critical as your organization grows. It is better to make the investment in these tools sooner rather than later. The longer you wait, the more difficult it becomes to implement. Delay creates problems with data migration, validation, and trending. Invest as early as possible and enjoy the benefits sooner.

5. Establish control mechanisms for the data collection process itself. Monitor timeliness of reporting, accuracy, errors and omissions in the data. Analyze the correlation between the lagging and leading metrics to see if there truly is a relationship. Monitor and review the process for data integrity and safety. Data integrity, safety, and privacy are important topics in the current environment and should be routinely evaluated. Make sure you have

appropriate controls to maintain the accuracy and consistency of data over its entire life cycle. Expect your metrics and dashboards to evolve as your organization grows, matures, and your business objectives change. But, have a plan for growth so you do not lose valuable historical data or the ability to understand trends and relationships. Assess the openness and transparency of information to management review.

There is an old definition for the acronym SMART, but it does not work well for QMS metrics. The following framework is better:

- **S**pecific—with a clear definition and criteria, established at relevant functions and layers
- **M**eaningful—tied to your quality objectives and quality policy
- **A**ttainable—including clear, reasonable goals with action levels and milestones
- **R**eal-time—metrics must demonstrate the performance of your QMS in real-time (or as close as possible)
- **T**ime-related—show metrics over time to observe trends, patterns, and show progress and improvement.

One of the challenges with metrics is being able to visualize status or compare/contrast multiple groups (e.g., manufacturing lines, plants, etc.) and being able to look at multiple factors as well. Some techniques that can be useful include:

- Goals, triggers, or action levels can be added to visualize magnitude of issues and define escalation criteria.
- Segmenting—it is important to segment data properly to understand details and pinpoint issues. Create scorecards that can visualize segments properly.
- Profiling—consider profiling or analyzing several factors together to create a "profile" of risk. Example a site that has an increasing trend of MDRs, repeat audit observations, and overdue CAPAs is showing that they are unable to address problems. Highlight profiles of risk in management review.
- Indexing—Indices (e.g., Consumer Price Index) can be useful to rollup information to higher levels of management review.
- Mapping—Data mapping can be useful to visualize data on two or three axes.
- Heat maps are graphical representations using color coding. They can help to visualize multiple risk factors.
- Color coding, dashboards, speedometers, and other visualization tools can make it easier to understand and interpret data. Many companies use green, yellow, and red color coding to highlight acceptability status.

Metrics should be "well dressed" when used for management review and decision making. Define expectations that SMART metrics have goals, supporting detailed data available for review, etc. Commentary is encouraged to explain the data. Commentary and an improvement plan are required for any red metrics.

ANALYZE

Once metrics are defined, they can be used to monitor performance, create improvement goals, and other business decisions. It takes skill, organization, and effort to analyze large volumes of data to turn it into useful information. We can consider analytics in the following categories (Fig. 8.3):

- Descriptive analytics help to describe the past and answer "what has happened?"
- Diagnostic analytics help to understand causes and answer "why did that happen?"
- Predictive analytics use statistical models and forecasting to understand the future and answer "what could happen?"
- Prescriptive analytics use simulation and optimization techniques to advise on possible outcomes and answer "what is the best option?" This is simply the next level of predictive analytics. It extends beyond predictive analytics by specifying which actions and criteria are necessary to achieve predicted outcomes.

Analytic capabilities are necessary to convert large amounts of data into useful information. Data can be characterized by:

- Volume—Analytic capabilities need to handle large volumes of data from transactional processes (e.g., complaints, CAPA, and batch records) and other sources.
- Variety—Analytics need to handle both structured and unstructured data. Unstructured data may need to be structured with metadata tags.
- Velocity—Analytics capabilities need to grow in order to handle an ever-increasing velocity of data.
- Variability—Inconsistency in the data can challenge the ability to analyze it.
- Veracity—The accuracy of data can vary greatly and affect the ability to analyze it.

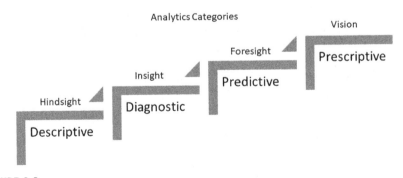

FIGURE 8.3

Analytics categories.

Both ISO 13485:2016 and the QSR require the use of appropriate statistical techniques. This applies to wherever data is used to monitor, control, and determine acceptability of process capability and product characteristics. It also applies to sampling plans, risk management, and improvement activities. Without appropriate statistical techniques all data is GIGO, garbage in and garbage out. Every medical device manufacturer needs to make sure they have the organizational capability for statistical techniques and include appropriately educated statisticians and/or certified quality engineers in the organizational structure.

Analytics must be able to show process stability, capability, trends, patterns, and issues. Beginner level companies resort to manual data capture and Excel spreadsheets to perform data analysis. They rely on structured sources of data and struggle with other types of data. They may transfer data from an automated system to spreadsheets for analysis. Manual manipulation of data lowers the veracity of information. More mature companies use more IT system data capture, advanced tools, statistical analysis, and business intelligence solutions. They are able to mine and analyze unstructured data sources. An even higher level of maturity allows the integration and networking of data from various sources (electronic document management system, product lifecycle management (PLM), laboratory information management systems (LIMS), complaints, CAPA, social media, sales, customer survey, etc.) for more meaningful information about the impact and value of quality and compliance. Increased analytics capability enables:

- Increased visibility of quality issues and impact
- Improved QMS efficiency by making it easier to track, trend, analyze, and report information real-time or near real-time
- Improved ability to measure process performance and improvement
- Enhanced quality and compliance due to a healthy QMS with real-time monitoring and control
- Improved risk prediction to define possible options including no action.

The potential to analyze and use multiple sources of data to gauge quality, customer sentiment, and market impact is ever growing and rapidly changing. We are now in the fourth industrial revolution and the world of "big data." This is a rapidly evolving field and medical device quality systems must rapidly change with it. Invest in data mining, data science, cloud computing, mobile apps, and connected sources of data. Be very mindful about data integrity and security. Medical device companies that embrace the opportunities of big data will be more competitive in the future. There is an ever-increasing amount of data available. Without good tools and analytics, it becomes noise that hides important information.

Analytics are necessary to turn large amounts of data into useful information. Priorities for analytics include:

- Real-time data availability
- Data quality and integrity
- Data stewardship and governance
- Accuracy, transparency, and openness.

Red flags to watch out for include:

- Manual data capture, analysis, and reporting results in accidental or deliberate errors, omissions, and inaccuracies of data.
- Unclear definitions, inconsistent application, or lack of valid statistical techniques corrupt ability to analyze, contrast, compare, and trend information.
- Metrics and data are presented in a misleading manner (altered scales, truncated data, missing or unexplained outliers).

Capabilities for monitor can be characterized in the following table:

Learning	Succeeding	Thriving
There is little to no use of metrics.	Metrics are required but are inconsistently used.	Metrics are consistently expected and used to make data-driven decisions.
	Metrics are used with some lag time.	Metrics are used in real-time or near real-time.
	Metrics come from disparate systems and data sources.	Metrics come from fully integrated systems to connect the dots, see correlations, and make judgments.
		Big data solutions allow additional sources and types of data enhancing business intelligence.
		Integrated product life-cycle metrics provide complete product quality intelligence.
Metrics are manually gathered, summarized, analyzed, charted.	Metrics are gathered, summarized, analyzed, charted with a hybrid system of automated and manual tools.	Metrics are all gathered and analyzed via integrated automated systems, allowing real-time analysis and action planning.
There is little or inconsistent use of valid statistical techniques.	Rudimentary awareness of statistical techniques characterized by frequent mistakes and assumptions (e.g. three points in a row going up is mistaken for a trend).	Statistical principles are used appropriately throughout the organization. Metrics and process monitoring data are presented correctly and enhance ease of understanding.
	Metrics are presented in charts and spreadsheets that are sometimes difficult for viewers to interpret and understand.	Metrics are presented using visualization techniques to enhance understanding. Metrics are backed up with more detail to support conclusions.

In conclusion, remember these basics when establishing your monitoring system:

- A consistent set of metrics and good analytics are prerequisites for understanding risks and prioritizing improvements.
- Good monitoring and measurement systems are essential to help you move from a mode of reaction to prevention.
- Regulators are constantly improving their abilities to collect, analyze, and act on data to protect the public health. Medical device companies would be wise to do the same.

FUTURE TRENDS FOR MONITOR, MEASURE, AND ANALYZE

The fourth Industrial Revolution is shaped by advances in connectivity and the internet of things. New technologies fusing digital and biological worlds, automation, and machine learning will allow improvements in manufacturing efficiency. Industrial leaders throughout the world are thinking about industry 4.0 and the opportunities it enables. This promises to affect not only the products that medical device manufacturers will be making in the future, but also our means to ensure quality and compliance. It is not too early for quality leaders to be thinking about the reality of Quality 4.0.

Radio Frequency Identification (RFID) technology allows tracking, traceability, and real-time information about manufacturing, distribution, and use. Automation in factories and more information about supply chain, receiving acceptance, in-process acceptance, finished goods acceptance, and customer use data will require a revolution in data analysis. Business intelligence, cloud computing, data mining, and mobile applications are all opportunities of the future.

Data analysis is critical to making good decisions. Data analysis in the future will need to address increases in data volume, velocity, and variety. Big data techniques will be important in the future to deal with huge increases in volume, velocity, and variety of data:

- Volume—Traditional Electronic Quality Management (EQMS) systems currently handle large volumes of data from transactional processes (e.g., complaints, CAPA, and batch records). In the future, they may be required to handle much more data from connected devices, sensors, and tracking. They will need to be integrated to understand relationships and connect the dots.
- Variety—Data will change from structured (e.g., complaint reporting and CAPA) to more unstructured data from connected products. Unstructured data may need to be structured with metadata tags.
- Velocity—Increases in the velocity of data should be expected in the future as technologies improve.

Connecting and analyzing information from fragmented and disparate IT tools (EQMS, Enterprise Resource Planning (ERP), PLM, LIMS, Supply Chain Management (SCM), and Customer Relationship Management (CRM)) will allow increased insights and predictive analytics. Fully connected and integrated systems will improve speed and accuracy to "connect the dots." Companies that do this well will position themselves for the future and be the most competitive.

EMBRACE

E is for Embrace. Embrace a culture of quality and compliance. Embrace values such as customer focus, prevention, transparency of information, and continuous improvement. Embrace data driven decisions and individual accountability. Embrace these ideals and make them the fabric of your culture.

Culture is the never-ending cycle where learned beliefs affect values that are manifested in behaviors. Shared behaviors create culture. Culture is complex, subtle, and sometimes difficult to measure and understand. And, culture has an enormous impact on quality and compliance. As you examine your capabilities with respect to culture, determine if your employees are aware of and committed to quality and compliance. Determine if they understand the quality policy and the impact of working in a highly regulated environment. Are they aware of the impact that they individually have on product or process quality? Are they trained and engaged? Do you have a culture that *embraces* quality and compliance? (See Fig. 8.4).

Organizational culture need not be random or accidental. It can be shaped by deliberate and consequential actions. It can be fed and encouraged to grow. Beliefs create the values that are articulated and shared by management. Beliefs and values are described in the quality policy and objectives, leading to personal objectives and performance measurement. This shapes improvement priorities, resourcing, and individual actions as well. These actions and behaviors affect the overall organizational culture. Individual actions become habits. Processes become more consistent with higher quality outputs. Collective habits become norms. And norms are the fabric of the culture. And the cycle continues.

Quality is not an act. It is a habit.

Aristotle

A culture of quality starts at the top. It is imperative that leadership is engaged and clearly articulates a vision of quality, compliance, and customer focus. Management needs to "walk the talk" to encourage a data-driven and transparent culture. Management demonstrates a passion for safe and effective products that help customers; that extend and improve people's lives. Management expresses a vision that quality is a competitive strength for medical device companies. Management demonstrates their belief that quality and compliance are part and

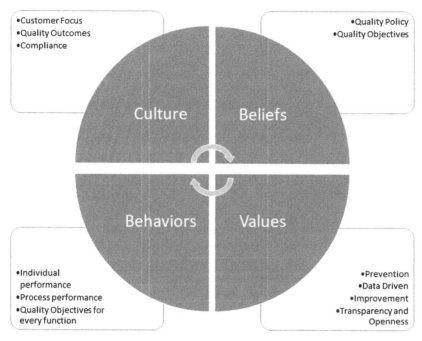

FIGURE 8.4

Culture of quality.

parcel of earning the trust and loyalty of customers. A credible message from management is a key factor in shaping the culture of quality.

TIP

Management should translate the clinical but dull regulatory requirements of "safe and effective" into more descriptive terms. A passion for excellence, helping customers, saving lives, reducing pain, improving lifestyle, reuniting families, and giving hope create a more compelling image. Some medical device companies do this very well.

Senior management and the quality council, through word and deed, shape a culture by defining company vision, strategy, and quality objectives. These become the beliefs and the values of the organization such as:

- Focus on the customer
- Quality and compliance are always top priority
- Do it right the first time
- Prevent errors and nonconformities
- Quality is everyone's responsibility
- Benchmark for new possibilities

- Drive continuous improvement
- Make data driven decisions
- Share information with transparency and openness.

Beliefs and values must be translated to the desired behaviors from the organization. Behaviors are a set of repeatable actions that are seen as acceptable and commonplace. They are seen as company norms. Changing norms takes deliberate and conscious effort. Unacceptable behaviors must be called out. Acceptable behaviors must be articulated and incentivized using recognition and rewards.

Top management incentivizes appropriate behaviors by setting quality objectives that are cascaded to every department, every process manager, every manager, and every individual. Quality objectives appear on the performance goals of every individual and are incentivized by performance measurement and compensation. With continued reinforcement, the desired behaviors become more consistent. Consistent behaviors of prevention and improvement create successes and reinforce beliefs. The cycle continues. With time, the organization will move to the state of unconscious competence, where these behaviors are so engrained and sustainable they simply feel like breathing.

The cycle requires authenticity from management. Employees easily recognize the difference between trite slogans and real management engagement. Employees can see when management is promoting quality because they read about it in a book and when management is doing it because it is the very essence of doing good business. Management must do more than talk about quality objectives. They must demonstrate it in every word they say, every decision they make, and every action they take!

Authenticity is the alignment of the head, mouth, heart, and feet—thinking, saying, feeling, and doing the same thing—consistently. This builds trust, and followers love leaders they can trust.

Lance Secretan

A culture of transparency and openness is incredibly important to acceptable quality and compliance results. Sharing accurate data and information is important in highlighting risks and priorities. Celebrate when issues are identified. It is an opportunity to improve. Known risks are always better than unknown risks and surprises! Use known issues to evaluate risk, prioritize activities, and identify needed resources.

WARNING

Be very aware of attempts in your organization to hide or downplay issues such as audit observations, quality problems, and overdue CAPAs. This is especially true when there is a profile of issues. I have seen some business managers go to great lengths to minimize very serious issues. In the short term, they may have gotten better performance evaluations or raises. Or they

(Continued)

WARNING (CONTINUED)

---

may have been able to focus on sales and product development for a while. In the long run, they ended up with a series of recalls, multiple Form 483s, warning letters, a recidivist warning letter, and ultimately a consent decree.

On the other hand, I've seen some business managers very openly and effectively communicate problems in a transparent, data-based manner and **use** that to justify the necessary resources (people, money, infrastructure, etc.) for improvement. This case study emphasizes the importance of a culture of quality, tolerance for risk, and ethical decision making.

Of course, management can just as easily drive the wrong values, behaviors, and culture. How many times have you heard comments such as:

- "We never have time to do it right, but we always have time to do it over."
- "We know what the real problem is, but they won't listen to us."
- "We told them about that months ago, but they ignored us."
- "I knew that was coming, but management didn't want to do anything about it."
- "We're always fighting fires and never doing anything positive."

The definition of insanity is doing the same thing over and over again and expecting a different result.

Einstein

A culture of quality can be characterized on several scales relevant to quality and compliance:

- Scale of prevention vs. reaction;
- Degree of collaboration
- Data driven vs. gut feel
- Degree of acceptable risk tolerance
- Degree of transparency of data and information
- Level of employee empowerment to raise concerns about quality and compliance
- Degree of process ownership, performance, and control
- Fear or willingness to share information factually
- Level of focus on the customer
- Degree of acceptance that everyone owns quality and compliance
- Degree of peer involvement and reinforcement of a culture of quality.

Culture can be reactive or preventive. In some companies there is always a fire that needs to be extinguished. People run around, make a lot of noise, and call emergency meetings disrupting other activities and consuming resources. These organizations are skilled at problem solving, but lack skills in process ownership, monitoring, and control. Reactive cultures may be moderately effective but do not optimize efficiency. Signs of reactive cultures that do not work well include:

- Audits are not conducted well.
- Audits do not really connect the dots.
- Problems and issues are not completely fixed.
- Improvements are not timely.
- Improvements are ineffective.
- Improvements are not sustained.
- Progress is not monitored.

Prevention is a key to **both** effectiveness and efficiency of your QMS. As such, it deserves special attention in a culture of quality. Prevention requires relentlessly standardizing and mistake-proofing processes, understanding cause(s) of errors and nonconformities, and preventing recurrence. Prevention of problems should be rewarded more than heroic fire-fighting efforts!

Culture can also be expressed in terms of compliance with policies and procedures. Is compliance seen as the norm, or does your company have a "get it done at all costs" mentality? The "get it done at all costs" mentality has resulted in many short-cuts, deviations, nonconformities, form 483 observations, and warning letters. It can also result in *qui tam*, or whistle blower, lawsuits under the False Claims Act. There are numerous medical device companies that have had to pay tens or even hundreds of millions of dollars to resolve False Claims Act violations.

The level of collaboration in an organization also has a strong impact on culture and results. Misalignment and lack of collaboration can result in covert and overt dysfunctional behaviors. People go in different directions and sometimes even work against each other. Process owners may neglect compliance and quality as they focus on cost and speed. Quality and compliance suffers. The quality organization compensates by overplaying the police officer role. Collaboration deteriorates even further. Lack of collaboration only increases the level of fire-fighting in an organization. It wastes precious time and resources.

CASE STUDY

I have seen organizations plagued by lack of collaboration. Leadership was unable to agree on even simple projects. They criticized each other behind closed doors and blocked projects that did not directly benefit their function. Or, they verbally agreed to projects but later sabotaged them with passive aggressive behaviors and lack of support. For example, they said they would provide a representative for an important project, but then, the representative was always too busy to attend team meetings. When the project team tried to proceed without complete team representation, absent members blocked the progress of the project. This behavior keeps the entire organization at the level of the weakest partner. The lack of collaboration flowed down through their subordinates. Ultimately results suffered, and the organization became disillusioned.

Data-driven organizations make better decisions. This culture starts at the top. A data-driven culture will use data to enhance quality system effectiveness and efficiency. It shows strong use of the metrics and analytics to identify, categorize, and prioritize risks. Management with executive responsibility, process owners, quality leaders, and even individuals can ask for and use data. Make data-driven thinking a part of an annual culture of quality survey. Create a common language of quality measurement and improvement (see Chapter 15: Alphabet Soup).

World-class quality and compliance requires a culture of process monitoring and control, process ownership, and prevention of problems. Various improvement methodologies are discussed in Chapter 15, Alphabet Soup, and can be used to create robust processes that are stable and in control. Stable, in-control, and capable processes result in better compliance and product quality. Capable processes yield more predictable results and enable business success.

Quality is everyone's responsibility.

W. Edwards Deming

Everyone owns quality. Employees must understand their role and contribution to quality and compliance. They must understand the relevance and importance of what they do. Until the entire organization embraces a culture of quality, the organization will be limited in their success. Plan, conduct, and act on a yearly survey to understand and improve your culture of quality and employee engagement.

Customer focus is a regulatory expectation in ISO 13485:2016. Clause 5.2 requires that "management shall ensure that customer requirements" are determined and met. Additionally, clause 8.2.1 requires organizations to "gather and monitor information relating to whether the organization has met customer requirements."

Obsess over customers.

Jeff Bezos

World-class companies focus on, indeed obsess about, the customer. They address all attributes of product and services. When thinking about quality, they think about the "Big Q" and not the "little q." They actively seek and react to customer feedback. They translate regulatory expectations for safe and effective products into more descriptive ideals for their products in order to drive deep awareness and commitment to customer needs.

Many organizations state quality objectives, but do not display an underlying commitment to them. It is important to periodically measure the commitment to and culture of quality in an organization. World-class companies make efforts to do annual surveys of all employees to determine the status of culture and engagement, and identify areas for improvement including:

- Is the company vision clear?
- Are quality objectives well defined?
- Is management viewed as credible and authentic?
- Are actions consistent with values and objectives?

- Is there a true focus on the customer?
- Are there consistent and appropriate quality decisions throughout the organization?
- Is information always shared in an open and transparent manner?
- Do employees feel free to raise issues?
- Do employees feel empowered to make quality decisions?

WARNING

Don't do a survey if you are not prepared to act on it. Walk the talk. You must share the results with the organization. This is the perfect opportunity to show that information is shared in a transparent manner. This is the perfect opportunity to show a commitment to continuous improvement.

The saying "culture eats strategy for breakfast" is often attributed to management guru, Peter Drucker. Although there is debate about who actually said it, it does emphasize the importance of culture. It does not mean that strategy is unimportant, rather that culture is a more powerful factor in business success. The best strategies and quality objectives in the world will be ineffective without a true understanding and focus on a culture of quality. In the medical device industry, the lives of patients depend on product quality, safety, and efficacy. In the medical device industry, culture eats strategy for breakfast, lunch, and dinner!

Companies that are beginners in shaping a culture of quality are reactive in nature and see quality as no more than trite slogans. Companies that are more mature make deliberate efforts to understand the culture and employee sentiment and attitudes toward quality. Management makes deliberate efforts to articulate the vision and quality objectives. Advanced companies have employees that all embrace the company's vision, values, and quality objectives. They unleash the potential for innovation and improved quality outcomes. Ethical and collaborative behaviors are the norm. Prevention is the standard expectation. Process ownership, individual accountability, continuous improvement, and pursuit of excellence are second nature.

WARNING

Be aware of the signs of significant immaturity in your culture of quality. Negligence is willful disregard for quality systems expectations. Obstructive behavior is deliberate action to avoid quality and compliance in an attempt to achieve schedule or cost goals. Contemptuous behavior occurs when quality and compliance issues are clear and there are deliberate attempts to disguise or cover up problems. Undermining occurs when individuals or functions downplay, criticize, or attack the actions of others to change priorities or compete for scarce resources.

Red flags to watch out for regarding a culture of quality include:

- Management commitment to quality is not credible. Management does not walk the talk. There is a disconnect between what they say and what they do.
  - They say they want openness and transparency but react negatively when they get it.
  - They say they make data driven decisions, but then revert back to "because I said so" mentality.
- Functional managers/process owners consider quality and compliance a bureaucratic exercise. They may even show signs of negligence, obstructive behavior, contemptuous behavior, or undermining actions.
- Management receives filtered, big-picture, "pretty" information from the organization and is unaware of true risks.
- Executive management does not value the input of the management representative.
- Employees are afraid to share to information or bad news.
- Employees feel frustrated by repetitive fire-fighting.
- There is a culture of fear and blame.
- A mentality of "get it done at all costs" overshadows the expectation of compliance.

Capabilities for embrace can be characterized in the following table:

Learning	Succeeding	Thriving
Lack of awareness regarding the culture of quality	Management with executive responsibility knows the Park Doctrine and is trained on the importance of a culture of quality	A culture of quality and compliance is seen as a business imperative
		Quality is first among equals—schedule and cost
	Company uses surveys to understand and improve the culture or quality	A formal, well-articulated program of quality-driven management exists and is actively emphasized
	Management with executive responsibility routinely articulates the importance of the customer, quality, and compliance	Management displays true engagement and authenticity for a culture of quality and compliance
The customer is considered through formal mechanisms like complaint handling	The customer is considered more broadly. More thorough systems such as design inputs exist to proactively seek customer input and feedback	The organization has shifted from a little q to Big Q mentality. Customer focus is a deep and enduring belief

(Continued)

Continued

Learning	Succeeding	Thriving
Little to no use of data, metrics, or information sharing	Metrics are used as a positive call for action There is sometimes fear in sharing information completely. Information is sometimes filtered to make it look more palatable	Openness, transparency, and candor are valued and expected competencies Metrics are standardized, shared accurately, and without filtering or fear of punishment Modern communication methods, real-time analytics, and integrated systems are embraced and used to drill down, connect the dots, and thoroughly understand risks in real-time
Decisions are made in silos or may be deferred to senior management	Decisions are made using evidence-based criteria	Decisions are made using evidence-based criteria. Stakeholder alignment and customer satisfaction are important criteria
Quality is seen as a department or a program and not as an individual responsibility	Quality is everyone's responsibility. All employees have quality objectives in their performance goals Every department has quality objectives in their performance metrics	All employees understand and embrace the importance and relevance of their work Processes exist to recognize and reward innovation that improves QMS efficiency and effectiveness Employee ownership and empowerment are defining characteristics Learning moments are constructed to share lessons learned, breakthroughs, best practices, and innovative improvements

In conclusion, remember the basics of a culture or quality and compliance:

- A culture of quality and compliance is not random. It is deliberately shaped and nurtured.
- Management needs to define values and quality objectives.
- Consistently demonstrate a commitment to quality and compliance. Walk the talk. Repeat again and again.

Give them quality. That is the best kind of advertising in the world.

Milton Hershey

FUTURE TRENDS FOR EMBRACE A CULTURE OF QUALITY

Leading companies will embrace the vision that compliance should result in improved quality outcomes. They will use their culture of quality as a trust mark to enhance their brand and increase the value of their products in the marketplace. They will use a culture of compliance and quality to earn customer loyalty.

DEFINE—RISKS AND PRIORITIES

D is for define risks and priorities. Risk management is an essential element of an effective QMS. Risk management is of strategic importance in understanding an organization's approach to quality, compliance, and customer focus. Companies encounter problems, issues, events, and/or nonconformities daily that require consistent and systematic evaluation to understand the nature of the problem and the associated risk. Proper risk management helps your organization assess, prioritize, and communicate issues systematically.

Risk tolerance drives the risks an organization is willing to take. It drives the decisions that are made. It drives priorities, and the amount of resources available. Regulators have seen the consequences of poor decisions and incorrect prioritization made by medical device companies. Because of this, the changes to ISO 13485 in 2016 placed a much stronger emphasis on risk. Risk is mentioned many times in the revised standard. Clause 4.1.2 of ISO 13485:2016 states that the organization "shall apply a risk-based approach to the control of the appropriate processes needed for the quality management system." Essentially this puts the burden on medical device companies to prioritize issues and allocate resources according to risk. Companies have no excuse for not addressing the high-risk issues.

Adding a risk-based approach to the regulations requires that medical device manufacturers define processes requiring risk-based decisions. Audits, CAPAs, supplier controls, and product risk management are common areas for risk-based decisions. Manufacturers should prepare justifications of risk levels and required controls for those various risk levels.

Additionally, risk management can be a way for companies to streamline their compliance in an objective and systematic way. It is essential for a highly efficient and effective QMS. Risk management requires a holistic approach including:

- Risk identification
- Risk evaluation

- Risk assessment tools
- Risk management decisions and priorities
- Improvement actions and resources
- Communication and escalation.

Risk assessment tools are necessary to help companies replace a "gut feel" approach to managing their QMS with a more objective, repeatable approach. Based on risk, a company can build a system of alerts, guidelines, decision criteria, and escalation of critical events. This enables the company to take actions that are commensurate with the risk.

Is risk management any different than preventive action within the context of CAPA? Some say that identifying and implementing preventive actions are essentially risk management. Certainly, risk is an element of CAPA. And yes, risk reduction can and should prevent problems. But CAPA does not cover the totality of risk management concepts in terms of assessment, decision making, resourcing, escalating, and communicating risks. It is more useful to think about risk management as an objective and systematic way to address risk holistically and use that to drive strategy, plans, and resources. In this section, we will look at broader concepts of quality and compliance risk within an organization.

Medical device manufacturers need an effective process for analyzing and categorizing risk. ISO 14971 is the recognized international standard for methodically analyzing, mitigating, controlling, and monitoring product risk. The concepts of ISO 14971 are certainly useful for a broader view of risk. But, we also need to think more broadly of the risks to your QMS and business. Risk management within the context of a QMS needs to address compliance risks as well. In this context, risk management helps to prioritize issues in order to make good decisions to ensure an effective QMS. Risk management is essential to making good decisions about how and where to allocate your resources for quality and compliance improvement.

Risk management concepts may be effectively used in CAPA and internal audit to understand risks and prioritize actions. Risk management needs to be integrated with your CAPA process, so that you address the most critical issues in a timely manner. This approach within the CAPA system is not meant to replace explicit risk management for the entire quality system, or risk management of your products.

WARNING

Do not go to the effort of categorizing CAPAs by risk and then treat them all the same with the same timelines and resources. Treating low-risk CAPAs this way will only dilute your resources and efforts to manage high-risk issues. Actions should be commensurate with risk. Take care of the highest risk issues first, but manage them with diligence, robust CAPA, and control plans to sustain the gains.

Effective risk management requires a consistent approach, taxonomy, and tools to properly categorize and classify risk. Risk is defined as the probability of **occurence** and the *severity* of harm. Risk can be considered in terms of:

- Risk to the patient
- Risk to the product
- Risk to compliance
- Risk to company personnel (safety)
- Risk to the environment
- Risk to the company (market share, financial impact, reputation, etc.).

Defining risk requires a consistent method of scoring risk. A risk matrix is an easy and useful tool to classify risk in terms of occurrence and severity. It is a powerful tool allowing consistent and repeatable analysis and classification of risk. Companies can use a risk matrix to make decisions about:

- Who is notified about issues
- Escalation of information
- Who needs to approve the CAPA
- Prioritization, timing, and allocation of resources.

The risk matrix generally is a grid of three to five occurrence categories and three to five severity categories. Some companies use a different grid for quality risk and compliance risk. For example, all new CAPAs can be evaluated in terms of product risk and compliance risk (see Fig. 8.5).

In the sample matrix above, items are coded as green, yellow, and red level risks.

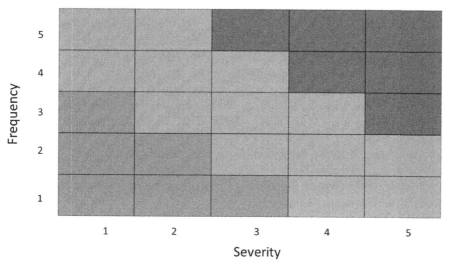

FIGURE 8.5

Risk matrix. Key: Green = low, yellow = medium, red = high risk.

One can analyze risk from the perspective of having a compliance issue, or event. An event can be minor on its own, or more severe when it occurs more often or is systemic throughout the organization. It may be an event that is significant enough that, if detected by external regulators, would result in a Form 483 observation (or major observation by notified bodies). It is important to use the risk matrix consistently throughout the organization for an "apples to apples" comparison. It needs to fit your unique company processes and product types. What is applicable to a company making high-volume, class I products may not work for a company making low-volume, class III products. Vet the risk matrix with real-life examples to make sure it fits your needs.

The risk matrix feeds decision trees about how to treat a specific issue. For example, all high-risk product quality issues should be escalated to VP level for review and approval. Decision trees can be diagrams in procedures or they can be built into IT systems. For example, all CAPAs classified as high risk may automatically require VP level approval of failure investigation, timelines, and activities. This is a perfect way to make sure that actions are commensurate with risks.

Another element of business risk management is safe writing and documentation practices including:

- Do not use inflammatory words in emails (for example "this is a disaster").
- Just stick to the facts and data. Do not use expressions like "If the FDA finds out about this, they'll shut us down."
- Be consistent with risk classifications and escalation triggers.

Risk management connects to the requirement for management review. It provides management with the information to make sound decisions regarding the suitability of the QMS. And it helps to ensure that **actions are commensurate with risk**. It is useful to consider risk in terms of known risks, unknown risks, and overall residual risk. This relationship is characterized in Fig. 8.6.

In order to reduce overall risk, management needs to be aware of known risks and mitigation actions. Known risks need to be characterized to determine scope, location, frequency, and severity. These risks need to be communicated in a clear and transparent manner. Mitigation activities need to be prioritized and resources allocated and managed to reduce the severity and occurrence of the risk. Improvement projects need to be monitored. Lack of resources and progress need to be escalated. Status of mitigation efforts needs to be communicated in management review.

And the unknown risks cannot be ignored. Management cannot avoid icebergs ahead by pretending that they are just not there! They must be located and course corrected. The only way to understand these unknown risks is to scan the external environment; continually monitor process performance and sources of data; and to continue to audit, making use of reasonable sample sizes and frequency. Identify areas that have insufficient information to understand risk (e.g., sites or processes that have not yet been audited). Predictive modeling (e.g., Monte Carlo simulations) can be used to understand possibilities and risks.

FIGURE 8.6

Residual risk equation.

Known Risks	Unknown Risks	Mitigate Risks
Characterize:	Scan environment	Prioritize
• Size, scope, location	Monitor sources of data	Provide resources
• Frequency	Continue to audit	Manage projects
• Severity	• Adjust frequency	Monitor progress
• Type (product, process, QMS)	• Increase sample size	Communicate status, progress, and adequacy of resources
• Systemic nature	Communicate areas of insufficient information	Escalate lack of resources or progress if necessary!
Communicate as applicable		
Escalate per risk classification matrix		

In summary, every medical device company needs to be aware of how they define, measure, and mitigate risks. A healthy QMS requires a conscious assessment of risk. Management with executive responsibility needs to take an active role in understanding risk as they review the suitability and effectiveness of the QMS. Actions need to be commensurate with risk. Resources need to be made available to reduce risks. Compliance and quality risk should be considered in all decisions from mergers and acquisitions, global markets, automation of quality systems, etc. In order to make compliance streamlined and efficient, broader thinking about risk is required.

As part of the organizational culture, every company has a tolerance to risk that must be consciously evaluated. Be aware of red flags that indicate unacceptable tolerance to risk in your organization:

- The organization deals superficially with corrections and corrective actions. Corrective actions only address the nonconformity immediately identified in the audit or CAPA. Failure investigation is not done to identify and remediate other similar nonconformities.
- Organizations that have a history of "getting away" with things will repeat bad behaviors, thinking it is low risk. Eventually they do not get away with it anymore. This destructive and pervasive attitude can bring down an entire organization. This is why the FDA issues recidivist warning letters.

- Some companies focus too much on new product launches without addressing known quality issues in previous generation products. It is an unacceptable risk to develop new generations of product without fixing the underlying quality issues of previous generations.
- False confidence causes organizations to dismiss facts and data that say something else:
  - For example, an organization dismisses a poor internal audit result with explanations of "We had an FDA inspection 2 years ago and didn't get a 483." Pay attention the stages of learning (see Chapter 7: Quality is not an Organization) for individuals and organizations that display false confidence.
- Managers attempt or take actions to prevent audits:
  - "Our notified body just reviewed that, so we don't need to do an audit." Remember, inspections are only a sample and a given point in time. Absence of regulatory observations does not mean that the process is robust. If another sample, or more current information, gives you additional information, do not ignore it.
  - "We should stop auditing because we already have so many things to fix." This will just keep the unknown from being known. It leads organizations right back to unconscious incompetence.

Capabilities for **D**efine risks and priorities can be characterized in the following table:

Learning	Succeeding	Thriving
There is little to no understanding of quality and compliance risk.	Product risk management concepts (ISO 14971) are used for product development.	Product risk management is consistent in premarket, production, and postmarket decisions.
		Product risk files, documents (Failure Modes and Effects Analysis, Health Hazard Analysis (HHA), etc.) are updated and kept current consistently using production monitoring and postproduction monitoring.
	Risk management concepts are limited to formal systems such as audit observations and CAPAs.	Risk management extends beyond ISO 14971 expectations for product risk.
	Although formal mechanisms exist for risk classification, they are not really used. CAPAs may be classified by risk, but all are essentially	Risk management concepts are used to inform data driven decisions in management review. Management with executive responsibility uses

(Continued)

Continued

Learning	Succeeding	Thriving
	treated with the same resourcing and management review.	risk concepts to ensure a suitable and effective QMS.
	The organization sometimes takes unacceptable business risks or shows too much tolerance for risk.	The organization consistently makes appropriate decisions about risk and allocates sufficient resources, as needed, to reduce risk to acceptable levels.
	Risk management is reactive and addresses known risks only.	Risk management is proactive, using multiple sources of data and predictive modeling. Prescriptive modeling can be used to evaluate "what if" scenarios.

It's your decisions, and not your conditions, that determine your destiny.

Tony Robbins

In conclusion, remember the basics of risk management within the context of an efficient and effective QMS:

- Know what all your risks and problems are. Keep monitoring and auditing to look for icebergs ahead.
- Evaluate and classify risks consistently.
- Do not take unreasonable risks.
- Fix the big stuff first. Make sure you have enough resources to do it well.

IDENTIFY

I is for identify. **I** is for internal audit. One of the key capabilities every medical device manufacturer must have is the ability to **self-identify** issues. Every medical device manufacturer needs to have excellent ability to self-identify their own quality and compliance issues. Those that cannot **self**-identify issues are doomed to be informed by external bodies such as regulators, notified bodies, dis-satisfied customers, and market sentiment.

There are many good books about the auditing process. ASQ (American Society for Quality) has a significant body of knowledge and resources, and provides certification of internal auditors and biomedical auditors. This book does not attempt to duplicate that body of knowledge. Rather, it will highlight some important lessons for medical device manufacturers. All internal auditors and their managers should be Certified Quality Auditors or Certified Biomedical Auditors to demonstrate their competence with this important body of knowledge.

Of course, medical device manufacturers are required by regulation to conduct quality audit. 21 CFR 820.22 requires that "each manufacturer shall establish and maintain procedures for quality audit and conduct such audits to assure that the quality system is in compliance with the established quality system requirements and to determine the effectiveness of the quality system. Quality audits shall be conducted by individuals that do not have direct responsibility for the matters being audited. Corrective action(s), including a reaudit of deficient matters, shall be taken when necessary. A report of the results of each quality audit, and reaudit(s) where taken, shall be made and such reports shall be reviewed by management having responsibility for the matters being audited. The dates and results of quality audits and reaudits shall be documented."

These requirements are important and can help to significantly improve a QMS. But internal audit, alone, is not enough to for a truly efficient and effective QMS. In Chapter 5, Waste and Inefficiency in a Quality Management System, the concept of the feedback cycle was introduced. Audit (and acceptance activities as well) are, by their very nature, lagging indicators and late in the feedback cycle. When issues are discovered they require not only corrective action but often significant rework and remediation. Self-identification shortens the feedback loop. The shorter the feedback loop, the more efficient and effective your identification process will be.

Therefore, we have a hierarchy of mechanisms to identify issues and nonconformities. Each company must define their own hierarchy to identify issues. Issues found higher in the hierarchy indicate problems that are more widespread and require more remediation (Fig. 8.7).

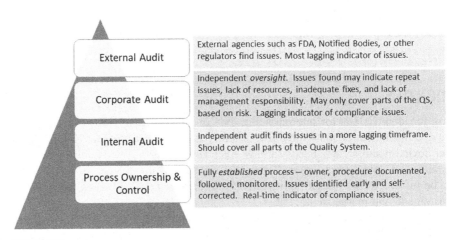

FIGURE 8.7

Identify hierarchy.

Not all companies have these many levels. Some may not have corporate audits. Some may have additional levels including site, business unit or franchise level, and shared stakeholder audits. But, each company should have a defined hierarchy. Issues found at one level indicate ineffectiveness at a lower level. Your risk classification determination should take this into account. Strive to identify issues as low as possible in the audit hierarchy. An organization that relies only on internal audit will always be reactionary in nature. To escape the endless cycle of reaction and rework, a company must develop proactive and preventive mechanisms to self-identify nonconformities.

The longer the time period from occurrence of a nonconformity to detection, the larger the volume of nonconformities that are created! This requires more correction, rework, and remediation. And there lies one of the significant problems with relying on internal audit to identify nonconformities. For a truly efficient QMS, you must have self-identification earlier than internal audit. Companies striving for true efficiency and effectiveness need more than internal audit.

TIP

Process ownership with process controls and self-inspection plays an important role in developing a more proactive approach to quality and compliance. Process ownership, process monitoring, and process control are critical to increasing not only effectiveness of your QMS but also **efficiency** of your QMS!

The following table describes a progression of techniques for identification from best to worst:

Process monitoring and control	Real-time process monitoring and control prevents issues form occurring and eliminates rework. Operators and individual contributors self-check their own work. Work is stopped if unexpected nonconformities occur.
Process checks	Process checks by the process owner occur quickly and drive prompt improvement.
Internal audit	A somewhat lagging indicator of issues. Nonconformities found here indicate problems with process control. Issues found here require closed loop CAPA.
Corporate audit	A more lagging indicator of issues. Nonconformities found here indicate that internal audit was ineffective in identifying issues; or that corrections and corrective actions were incomplete, ineffective, late, and did not address remediation of other instances. Since this is more lagging, there will be more rework or remediation required. Responses to Corporate Audit observations should have the same rigor as external inspections.
External inspection	External inspections may occur months or even years after issues occurred. Correction, corrective action, remediation activities for Form 483 observations, and warning letters can be staggering.

(Continued)

Continued

Customer feedback	The customer is the final authority. Customer complaints are an important indicator about the health of your QMS. Class I recalls, warning letters, and consent decrees reach the evening news. Issues can result in Medical Device Reports (MDRs), corrections and removals (recalls), loss of good will, and loss of market share.

As discussed in Chapter 2, Increasing Expectations, regulatory expectations for medical device manufacturers are ever increasing. One very important change discussed was the FDA's "Case for Quality" creating the expectation that compliance alone is not enough. Compliance should result in improved quality outcomes. This important concept needs to stay front and center in a good internal audit program. An internal audit program must be much more substantial than finding minor GDP (Good Documentation Practices) errors. A good internal audit program can identify real issues, connect the dots, and paint a clear picture of risk. A good internal audit program is an important part of getting to improved quality outcomes.

A sign of maturity in your audit program is when audits are approached with openness and transparency. No functional manager or process owner really likes to hear about problems with their process. But, a quest for excellence and accountability must be a stronger force. When this occurs, opportunities for improvement are welcomed and acted on. Defensiveness is not displayed. Audits are approached as an opportunity to make unknown risks be known. And it is *always* better to know about issues than to be surprised during external regulatory inspections. The organization or individuals should NOT be punished for receiving audit observations.

However, repeat audit observations are a real concern. Repeat audit observations indicate that management has failed to address known problems within the organization. This can be due to negligence, obstruction, or contemptuous behavior. This violates the basic concept of management responsibility. Repeat observations should be always be communicated and escalated to the next level of management for review and analysis.

WARNING

Because repeat observations are escalated to the next level of management, be prepared for push-back (sometimes significant) from the responsible party. They will try to justify why the repeat is different. They will claim it was a different product, or a different machine, or a different operator and therefore not really a repeat. They may spend more time arguing about the observation than they do fixing it. They may engage in undermining behavior. They may even spend time attacking the auditor and the audit process. They may claim they are being treated unfairly. This behavior is very destructive and only emphasizes the need for robust, comprehensive correction, corrective action, and preventive action.

Management with executive responsibility should be aware of different opinions regarding repeat observations. Failure to heed the warning signs can be very costly!

Regulators take repeat issues seriously too. When they do follow-up inspections, they will review past issues. Repeat issues are a key factor in escalation from a Form 483, to a warning letter, recidivist warning letter, and eventually consent decree. Regulators will attempt to connect violations with individuals that have a duty, power, and responsibility to detect, prevent, and correct violations.

WARNING

Determination of culpability is not an expectation for internal audit. But, internal auditors should be very sensitive to indications that someone acted improperly or failed to act properly. They should be alert for indications that this caused or contributed to issues that would be considered violative by the FDA. All company auditors should be alert for policies, practices, and politics that inhibit an effective QMS and compliance with regulations. These include:

- There are signals that the quality organization or management representative does not have real authority.
- Authority is not commensurate with responsibility.
- Personnel do not have clear roles and responsibilities.
- There is competition for scarce or inadequate resources.
- There is insufficient management support or responsibilities.

These concerns should be escalated and reviewed with higher level management. Additionally, process owners and auditors must be aware of any data integrity issues. Data integrity issues, whether accidental or deliberate, should always be taken seriously. Data integrity may be compromised by error, incompetence, overwork, or distraction. Any suspected falsified documents should be copied, isolated, and controlled as soon as possible after discovery. Evidence of deliberate falsification or destruction of records (fraud) must be carefully documented and escalated (including legal and human resources).

Another reason that process ownership and control is important is that internal audit may not find every single issue. It may not find every single example of an issue. Remember, an audit is only a "snap-shot" in time. It is a sample. It is NOT 100% inspection (and we know 100% inspection is not 100% effective). Again, we must rely on process control to provide a more comprehensive and complete picture of process performance.

A single audit, on its own, may not be able to connect all the dots. But, it can provide useful information. The information must drive useful improvements. And it must be well-communicated to those with the authority, power, and responsibility to address issues and provide adequate resources. A successful audit program should consider risk areas, communication of issues and associated risk, and result in reduction of risk. Build up and summarize your history of audit information. This information, along with other QMS metrics, paints a more complete picture of QMS performance.

Have a robust plan for where to audit. The approach should not be "ready, fire, aim" but "ready, aim, fire." Internal audit will need to cover the entire QMS while corporate audit may only cover key risk areas. Even though internal audit will need to cover the entire QMS, they do not have to cover it equally. You can schedule high-risk areas earlier in the audit schedule. You can cover high-risk areas in greater depth. Do a risk assessment every year to maximize your effectiveness. For corporate audit, use a risk-based approach (like the FDA does with "For Cause" Inspections, see Chapter 16: FDA Inspection Readiness). Consider risk factors such as:

- Who in the hierarchy is finding issues?
- Have past audits been addressed with transparency and openness?
- Is there a history of repeat observations?
- Are issues isolated or systemic?
- Determine criticality of issues and impact on product. Significant quality issues indicate problems within the QMS. They are often the starting point for FDA inspections.
- Look at other performance metrics to see if CAPAs, complaints, etc. are being addressed in a timely and effective manner. If one of your sites has a profile of issues, you can prioritize that site to put it earlier in your audit schedule, or to spend more time there.
- Are there any other changes of risks such as:
  - Changes in manufacturing location
  - Changes in suppliers
  - Acquisitions
  - New product releases
  - Product design changes
  - Management changes
  - Any changes in regulations or regulatory focus.

Keep a database of risk factors for each of the entities covered by your audit program. Include risk factors such as:

- Dates and results of external inspections. The FDA attempts to inspect all manufacturers of Classes II and III devices every 2 years. If it is been longer than that, the site is at risk of an inspection.
- Significant quality issues such as an increase in MDRs or a Class I recall may trigger and FDA inspection.
- New Class III products will trigger pre- and post-premarket approval (PMA) inspections.
- High-risk devices, or certain focus areas prioritized by the FDA (e.g., infusion pumps), may prompt inspections.
- A previous inspection with poor results (OAI or official action indicated) may have a follow-up inspection.
- New management or management that has previously displayed red-flag behaviors.

Create a heat map (see Chapter 16) of inspection likeliness versus potential severity and use it to prioritize your limited resources. Plan your program. Plan your approach depending on the factors identified in the database and on your heat map. It is possible to use a top-down approach that starts with the looking at the quality system and associated procedures. A bottoms-up approach starts by looking at one or more individual problems that point to a failure in the quality systems. It is also possible to audit a product through the entire product lifecycle; or conduct a horizontal audit looking at connected processes working together.

World-class quality systems are highly dependent on world-class auditors. They require world-class auditors that can effectively find and document meaningful audit observations. Auditors need to be able to consistently apply risk rankings, so that issues can be appropriately prioritized later. They need to provide linkage to the QSRs and to your own QMS. Auditors need to be able monitor improvement and document repeat nonconformities. Good auditors can connect the dots to turn noisy data into actionable information.

World-class auditors are the eyes and ears of your QMS. They are informed about the external environment and changes in regulation. They understand the focus areas of the FDA and other regulators. They are aware of the Form 483 and warning letters that your competitors or suppliers have received. Make a concerted effort to keep your auditors informed about external and internal issues and changes. Make sure your auditors are informed about management review, metrics and dashboards, key risks, and improvement actions. Do not keep your auditors in the dark and then expect them to be enlightened auditors.

George Bernard Shaw infamously said, "Those that can; do; those that can't, teach." This unflattering viewpoint is sometimes applied to auditors as well. People that are ineffective in other disciplines are often "put out to pasture" or relegated to the role of auditor. This is a huge mistake for a medical device manufacturer. It results in meaningless and annoying audit observations. The organization then reacts with bureaucratic responses while real risks go unidentified. A more appropriate ideal for auditors is:

Better me than a 483

Auditors need to have the following skills to provide meaningful audit observations that can prevent Form 483 observations:

- QMS expertise to understand regulatory expectations and best practices
- Subject matter expertise for the area being audited (e.g., design control, sterilization, and software)
- Technical skills, capabilities, and experience (e.g., understand product or process technology)
- Interpersonal and communication skills are necessary to interact effectively with auditees and stakeholders
- Good writing, reporting, follow-up skills
- Be trusted, valued, respected.

This is a significant skill set that is difficult to find. Invest in getting the best auditors. Create a development program for auditors (e.g., see one, do one or X, lead one). Require external ASQ certification. Rotate high-potential personnel through auditor roles. Identify and cultivate subject matter experts that can work with auditors to enhance key technical areas (e.g., sterilization). Have junior auditors work under more senior auditors to develop their skill sets. Bring in external consultants as needed to complement internal skill sets.

Keep auditors informed of internal issues and priorities. Let them have visibility to management review. Allow them to go to occasional outside conferences to understand industry changes. Make sure that they have visibility to changes in the external environment or regulations. Do not keep auditors in the dark and expect them to be excellent auditors.

Additionally, it is important to make sure that auditors and auditor management are independent (do not have direct responsibility for the matters being audited) and are not punished for pointing out problems. Maintaining independence is essential for ensuring that auditors can openly and clearly document issues. They should NOT be punished for pointing out problems! Watch for signs of undermining of auditors. Do not contaminate auditors by making them responsible for taking corrective and preventive action. This will infect your entire audit program.

WARNING

Do not blame the auditors when corrective actions are not taken appropriately! The auditees/process owners and functional managers need to be held accountable for process performance. The process owners and functional groups need to be held accountable for CAPA to address audit observations. Blaming the auditors not only takes accountability from where it belongs, it irreparably damages the effectiveness of the audit process. Auditors lose independence and objectivity. They don't want to identify issues they will be saddled with and know they cannot possibly fix. They become demoralized when undermined and lose the will necessary to connect the dots and identify the big risks. Auditor performance then declines and results in focusing more and more on small GDP details only.

Determine the impact or criticality of audit observations individually and as a whole. It can be helpful to calibrate against FDA considerations. The FDA compliance guidance program establishes two categories called situations I and II. Situation II is the least significant and is defined as QS/GMP (good manufacturing procedures) deficiencies of a quantity and/or type to conclude there in minimal probability, in light of the relationship between quality system deficiencies observed and the particular product and manufacturing processes involved, that the establishment will produce nonconforming and/or defective finished devices. Situation I includes documented evidence of one or more major deficiencies. Major deficiencies include:

- Total failure to define, document, or implement a quality system of one of the seven subsystems
- A deficiency in one or more element(s) of the subsystems

- The existence of products which clearly do not comply with the manufacturer's specifications and or the QSR and which were not adequately addresses by the CAPA program
- Noncorrection of deficiencies from previous inspection(s).

The greater the actual or potential impact on patient safety, the more significant the observations. Audit observations that demonstrate direct adverse impact on product quality are always more significant than those that pose potential.

Data integrity errors are usually not significant unless they occur repeatedly or there is a pattern to one area. Instances of nondeliberate errors that result in raw data that does not agree with informal records are considered serious. Instances of deliberate data falsification, unauthorized changes to data, obscuring or obliterating data, discarding or hiding adverse data should be considered extremely serious.

Isolated instances of nonconformity are generally less concerning than multiple instances (although some isolated instances can be very serious on their own if they have patient safety or product quality impact, or they are deliberate). The period of time from when a nonconformity occurred may also be a factor. If issues are old and corrections have taken place, they will be considered less impactful. Observations that occurred long ago but not recently are not as serious as current issues. Patterns and trends should be noted. Audit observations found higher in the audit hierarchy should be viewed with more concern and can be given additional weight.

The grading system used for Medical Device Single Audit Program (MDSAP) audits uses the Global Harmonization Task Force (GHTF) document *QMS— Medical devices—Nonconformity Grading System for Regulatory Purposes and Information Exchange.* This document uses similar concepts of impact on product safety and the nature of a first or repeat observation, but it structures it differently. This document provides a useful structure based on QMS impact and occurrence. Impact is considered as indirect or direct. Occurrence is considered as a first or repeat occurrence. The document provides clear rules on grading and escalation.

Medical device manufacturers should use these concepts to create a classification scheme for internal audit observations. The following table is an example of a potential audit observation classification scheme. Make sure that you customize it for your specific organization structure, product risk levels, and audit hierarchy.

Risk Level	Compliance Definition	Product Definition	Action
Critical	Nonconformity has potential or known impact on product safety.	Likely or possible to result in hazardous or unsafe conditions or prevent performance of a vital product function.	CAPA and correction.

(Continued)

Continued

Risk Level	Compliance Definition	Product Definition	Action
Major	Nonconformity that is systemic in nature.	Likely or possible to reduce product function and/or result in customer satisfaction.	CAPA and correction.
Minor	Nonconformity that appears isolated based on audit sample size.		CAPA and correction. CAPA not necessary with appropriate rationale.
Opportunity	Opportunity to apply a best practice or make process more consistent.		Determined by process owner.

Repeat observations are assigned one severity level higher. Overall audit classification is based on all of the individual observations and classification:

Audit result—critical:
- Audit identifies one or more major nonconformities in an element
- Repeat instances of major audit observations from last audit
- Systemic failure of the quality system or one of its subsystems
- Data integrity issues that appear deliberate.

Audit result—major:
- A number of minor nonconformities in a given element
- Repeat instances of minor audit observations from last audit
- Failure of one of the subsystems that is not systemic
- Data integrity issues that impact records.

Audit result—minor:
- No or isolated lapses in execution.

Another key concept to apply to internal audit is the need for robust, comprehensive improvement in response to identification of issues.

WARNING, WARNING, WARNING

One of the key problems with internal audit is that although the organization becomes aware of issues through internal audit, they do not put comprehensive correction, corrective action, remediation, or preventive action in place. The risk remains and is later identified by external regulators. Companies that do not effectively deal with issues are condemned to getting Form 483 observations and warning letters. In my work with clients, I find that approximately 75% of Form 483 observations are nonconformities that have previously been identified in internal audit. This is not only ineffective but inefficient and results in more remediation and more rework.

> **CASE STUDY**
>
> One of my clients received a significant warning letter from the FDA. In helping them address the warning letter, I became aware that all of the issues had been previously identified through internal audit. The internal auditors, who were helping to write the warning letter response, were very discouraged that internal audit had previously identified and documented **all** the issues in the warning letter. Due to the warning letter, PMA approvals were put on hold and the company could not launch an important new product. This company spent 2 years and tens of millions of dollars dealing with this warning letter.

Be aware of management response to internal audit. Sometimes management displays a lack of interest or concern and does not attend the audit closing meetings. Management sometimes does not read audit reports, audit responses, or monitor corrective actions. This is negligence of responsibility.

> **TIP**
>
> This is a significant risk factor that should be documented in your risk assessment. Consider how to escalate this concern to higher level management.

Because of this, CAPAs are often late, incomplete, ineffective, or not sustained. This is a primary cause of an inefficient and ineffective QMS. And, sadly, it is common. You must ensure that response plans and CAPAs for audit observations are:

- Timely: Timeliness should be commensurate with risk. Audit observations that identify product quality problems must be treated with appropriate urgency.
- Complete and commensurate with risk. They should address:
  - Mitigation to lessen the impact of a nonconformity (includes corrections and removals if product is in the field)
  - Correction of the nonconformity identified in the audit observations
  - Root cause analysis to determine root cause(s) and contributing cause(s)
  - Corrective action to eliminate cause(s);
  - Remediation to identify other instances of the nonconformity and address historical or legacy documents
  - Effectiveness checks to ensure that actions accomplish their intended purpose.
- Thorough:
  - For critical observations, ensure containment of nonconforming product including health hazard evaluation, product hold, and corrections and removals.
  - Investigate for more instances of the nonconformity. Nonconformities identified in the audit are based on a small sample. Review a larger sample size to identify similar instances and a more accurate failure rate. Clearly define the scope and extent of the problem.

- Is root cause analysis thorough? Is it due to process capability and control, procedure or documentation, lack of execution to the procedure, or other issues? Use the 5 Whys technique to understand cause(s) and contributing cause(s).
- For process issues, is the process stable and in control? Is it capable of meeting requirements?
- For procedural issues, also answer why the procedure was incorrect. Are there any other procedures, manufacturing lines, departments, or sites with the same issues?
- For execution issues, who needs to be retrained and why? How will you control training for personnel in the future? Can you mistake-proof the process?
- For remediation, identify the population affected (number of records, files, batches, etc.) and what portion will be remediated. How far back will you go to address old records or other instances? What is the rationale? What is the residual risk?
- For repeat observations, what is the remediation plan?
- What process checks or control plans will be put in place to "sustain the gain"?
- Does effectiveness check truly address the root cause(s)? Does it include an adequate sample size? Is it realistic?
- Well communicated:
  - Has the process owner created the response and communicated with Quality and appropriate management?
  - Are timelines and action items well defined?
  - Are resources available and in place?
  - Do any issues need to be escalated?

A sign of QMS maturity is that organizations are open to feedback, audit observations, and opportunities for improvement. There is an attitude that we are all on the same team. Process owners with a high degree of maturity welcome input and the opportunity to make their processes even more robust. They use problems to justify resources for improvement.

There is often confusion if records of internal audit and management may be reviewed by the FDA. Internal audit reports and management review are listed as exceptions to the general requirements for making records available to the FDA per section 820.180(c) of the QSR. However, upon request of the FDA, an "employee in management with executive responsibility shall certify in writing that management reviews and quality audits required under this part, and supplier audits, where applicable, have been performed and documented, the dates on which they were performed, and that any required corrective action has been undertaken." Further, comment 166 from the preamble to the QSR states "Two comments stated that the records required under this part be treated as part of the internal audit. FDA disagrees with these comments because this information is directly relevant to the safety and effectiveness if finished medical devices. FDA has the authority to review such records and the obligation to do so to protect the public health ...Manufacturers will be required to make this information readily available to an FDA investigator. ..." So, you do not have to show the FDA your entire audit or management review reports. But, you will be required to show evidence that you are doing internal audit and management review and that you are taking appropriate actions in response to those activities.

WARNING

Be concerned about efforts to keep audit observations and resulting corrective action separate from other CAPAs. Some organizations make big efforts to keep audit observations and resulting corrective action in a separate data base or system in an effort to keep them out of view of FDA investigators. More often than not, this backfires.

Additionally, when audit nonconformities are kept separate from CAPA nonconformities, it is impossible to understand risks comprehensively. Two separate sets of books do NOT facilitate good risk management and prioritization of resources. Bottom line, it is more important to focus on doing those activities in an efficient/effective manner than it is trying to hide them from your CAPA system.

Audits should not be seen as a negative thing but as a routine self-awareness exercise. Auditees that have high self-awareness should not be surprised by the results. They should have a good sense of what the issues are. And they consider feedback as a valuable element of improvement.

Audit your audit program. If you do not have a corporate audit program, then you should use an external third party to audit your internal audit program. Make sure that you can pass the red-face test. Make sure you have a true picture of the maturity of your internal audit process. Your success depends on it!

Watch out for red flags:

- Be watchful for excessive arguing from auditees about audit observations and risk level. Arguments and comments like "Where does it say exactly that in the regulation?" or "Why is that a critical observation?" indicate that something is wrong with your audit process. It can be unskilled auditors writing poor observations. But it can also be auditees that are fearful of having audit observations and problems known by higher management. Often times, it means that auditees acknowledge there is an issue, but are willing to take unacceptable risks. Failure to understand the reason can present unacceptable risk to your organization. It is important to determine the cause. At an extreme, this behavior is undermining of auditors and the audit process.
- Avoidance or objections to being audited:
  - "We just had a notified body audit. We don't need an internal audit."
  - "We already know what's going on."
  - "We're too busy and don't have time for an audit this year. We need to get product XYZ launched."

- Attempts to downplay or minimize observations:
  - "That's not really a repeat observation. It's different because of *XXXX* (some minute difference). We don't need to tell management about it."
  - "We had an FDA inspection 3 years ago and didn't get a Form 483." Remember, inspections are only a sample in time. A different sample on a different day may show issues. And 3 years ago is a long time!
  - "We were just audited by the Notified Body, and *they* didn't write it up."
- False confidence and unacceptable risk tolerance:
  - A history of issues and getting away with it creates a false sense of confidence. Superficial fixes and recidivist behavior will be noticed by the FDA and dealt with increasing levels of enforcement!
  - Companies sometimes take unacceptable risks such as designing a Class III second generation product, based on a predicate device with unacceptable product quality, and not addressing the quality issues.
- Trends and patterns that indicate systemic issues:
  - Audit observations for one auditee across multiple QS elements. For example: A site with multiple observations in complaint handling, late MDRs, and overdue product CAPAs may have problems dealing with product quality issues.
  - Systemic issues across multiple auditees may indicate systemic process issues or issues with IT enabled management systems.
  - A pattern or multiple observations in one QS area indicates an out of control process.
- Repeat audit observations are a big, waving, flapping red flag.

The capability to identify is characterized in the following table:

Learning	Succeeding	Thriving
Uses internal audit as the only method of self-awareness.	Adds oversight of the internal audit process. Adds additional approaches, bottoms-up, horizontal audits.	Incorporates preventive approaches. Uses real-time process monitoring and process controls.
Identifies GDP issues and isolated nonconformities.	Creates a more comprehensive analysis of risk using profiling and other techniques.	Systematic analysis of risk and approaches to "connect the dots."
Little evaluation of risk.	Uses risk concepts to prioritize audit areas.	Uses predictive approaches to improve compliance.
Focus is on correcting only the issues identified in the audit observation.	Focus is on process Improvements and comprehensive CAPAs. Includes remediation activities.	Focus is on process stability, capability, and control.
	Focus is on reducing repeat observations.	Focus is on improved quality outcomes.

(Continued)

Continued

Learning	Succeeding	Thriving
Internal audit viewed as cumbersome and burdensome.	Internal audit is seen as a value-added mechanism for self-evaluation.	Internal audit is seen as a double check to ensure proper process performance and control. Audit observations are rare and minor, but welcomed as additional insight and an opportunity to improve and strive for excellence.
Auditors are seen as nit-picky and nonvalue added.	Auditors are seen as value-added contributors.	Auditors are seen as respected experts adding insight to process performance and control.

In conclusion, remember the basics for **I**dentify:

- It is always better to know about issues than to be surprised.
- Shorten the feedback loop. It is always better to know about things sooner rather than later.
- Fix the things you do know about.
- Keep auditing to understand the things you do not know about.
- Remember, companies that do not effectively self-identify issues are doomed to be notified by regulators and customers. That is the height of ineffectiveness.

CAPA and Improvement

C is for CAPA and Improvement. CAPA is a formal closed-loop approach to improvement. It is a key regulatory requirement. As such, it is always part of an FDA inspection. An effective CAPA system is essential to an efficient and effective QMS. But CAPA, alone, cannot cure all ailments. CAPA is but one of several necessary processes. Indeed, ISO 13485:2016 places CAPA under Clause 8.5 Improvement.

Most organizations struggle with real improvement. They fail to understand the distinctions between correction, corrective action, and preventive action. They limit improvement to only the specific example (nonconformity) that they know about without looking more broadly. CAPA is usually limited to correction only. Companies fail to understand the true scope and magnitude of the issue and identify other instances. And so, the problems keep coming back again and again. Companies that are already resource constrained waste precious time and money without getting better results.

Those that fail to learn from history, are doomed to repeat it.

Winston Churchill

There are better ways to do things. Learn from mistakes. Share learning. Because improvement and CAPA are such large topics, an entire chapter is dedicated to them. See Chapter 15, Alphabet Soup, for additional detail. We will give an introduction in this section to understand CAPA and Improvement as a key capability necessary for an efficient and effective QMS.

A key to meaningful improvement is knowing the difference between correction, corrective action, and preventive action.

- Correction—the official definition is "action to eliminate a detected nonconformity." This means that you have already identified a nonconformity or problem. This can be nonconforming material on the manufacturing floor or a complaint from a customer. You take action to fix this known nonconformity. You must investigate to fully understand the scope and magnitude of the nonconformity and see if there are more nonconformities to correct.
- Corrective action—the official definition is "action to eliminate the cause of a detected nonconformity or other undesirable situation. Corrective action is taken to prevent recurrence." This means that you understand the cause(s) of the nonconformity. You eliminate the cause(s) to prevent the nonconformity from happening again.
- Preventive action—the official definition is "action to eliminate the cause of a potential nonconformity or otherwise undesirable situation. Preventive action is taken to prevent occurrence." This means that a nonconformity has not yet occurred. It means that you are monitoring your processes and products on an ongoing basis. You are continuously monitoring these sources of data for process stability and capability. You are looking for changes, patterns, or trends to see what is happening. You take action before a nonconformity occurs.

A preventive action is not necessarily tied to every corrective action. The GHTF guidance document on CAPA provides the best explanation. It states that "the acronym CAPA will not be used in this document because the concept of corrective action and preventive action has been incorrectly interpreted to assume that a preventive action is required for every corrective action. This document will discuss the escalation process from reactive sources which will be corrective in nature and other proactive sources which will be preventive in nature." So companies must have proactive sources of information. Once again, this ties into the process approach to QMSs. By monitoring process performance and other sources of quality data proactively, it is possible to see trends, patterns, and out of control conditions that indicate preventive actions are necessary.

Most companies are bad at preventive action. They are bad at preventive action for many reasons:

- Sometimes companies are bad at preventive action because they do not proactively monitor sources of data in a meaningful manner. They do not recognize the patterns, trends, or conditions that indicate something is changing. They are not skilled or aware enough to recognize "potential nonconformities." They cannot prevent problems they have no idea are coming.

- Sometimes companies are bad at preventive action because their resources are already consumed with corrective action. They stay in a perpetual mode of crisis management and firefighting. Until they can break even a small portion of their resources away from reaction and move them to prevention, they will stay in the crisis—management cycle.

- Other times, companies recognize a potential issue but do not want to put preventive actions into the formal, closed-loop CAPA system because they think it is too burdensome. They think that they can take care of preventive actions on the side, without the rigor of a closed loop CAPA system. But, that makes no sense. Serious issues that must be prevented are exactly the ones that **need** the rigor and robust problem-solving methods of a formal CAPA. The potential for serious issues requires the attention of management that CAPAs should get. Potential high-risk issues that are preventive in nature should go through your formal CAPA process.

Of course, preventive action, in a broader sense, can include all of the many things you do on a regular basis to ensure good quality and compliance (reference Cost of Quality Model in Chapter 5: Waste and Inefficiency in a Quality Management System):

- Training
- Process qualification and validation
- Quality planning
- Design control
- Risk management
- Customer feedback
- Thorough, meaningful, and complete responses to audit observations.

Well then, the entire QMS is about preventing problems. Quality problems can always be traced back to quality system problems. Remember to ask why five times (see Chapter 15: Alphabet Soup) when you have quality problems, so that you can understand why your QMS failed. Tie failures back to immaturity of one of the MEDICS.

Many companies fail to investigate to understand the full scope and magnitude of issues. They take the easy way out and only address the issue right in front of them. They fail to look further and see if there are any additional instances of the same or similar problem. And they fail to address all instances of the issue.

High-risk issues and complex problems need good project management to ensure progress. Project management is useful to make sure that improvement activities are complete, appropriate, timely, and monitored. Project management expertise and tools can add thoroughness, consistency, and improved records to CAPAs. Big problems sometimes require a team of people to address them. A project manager has the tools to organize and lead a team to better success.

High-risk CAPAs require effective project management to ensure improvements are made in a prompt, coordinated manner. Remember, actions must be commensurate with risk. Consider it an investment to add a project manager to support, or even lead, high-risk CAPAs.

Improvement also requires change. Change can be intimidating and threatening to individuals and organizations. Change is difficult. It requires work. Because of this, organizations or individuals resist change. This can be especially true of quality and compliance improvements that are seen to cut headcount and resources. Be aware of the warning signs. Use change management tools to improve your success.

There is no need to join the flavor of the month club and switch from one miracle improvement methodology to another. What is more important is that you have a consistent problem-solving methodology, a good tool box of tools, and a common dictionary to speak the same language. Management that consistently asks for data, next steps, and goals will institutionalize the expectations for real and lasting improvements.

Improvements for audit observations must be complete, accurate, and thorough. This is the heart of the matter. Effectiveness checks must be realistic. They must address the root cause(s) and be measurable. Predetermined criteria for success are important. Criteria for success should be reasonable—not too easy or too hard to meet. If it is too easy, you will not know if you have really achieved the desired level of improvement. If it is too hard, you will fail and need to start over. If you have done a good job at verifying and validating corrective action, then your criteria for success should be realistic.

WARNING

If you fail to meet criteria for success, do not rationalize why that is acceptable and then close the CAPA as effective. A failed effectiveness check means that proper homework and preparation have not been done (adequately analyzed root cause and completed acceptable verification and validation). If you fail an effectiveness check, it is important to go back to root cause analysis.

ANOTHER WARNING

If you fail to meet an effectiveness check, do not close the CAPA and open a new one without any connection to the initial CAPA. I have seen companies do this two, or even three, times always making each CAPA seem like a new and minor issue. This is particularly problematic for critical product quality problems! Other companies have CAPAs that fall behind schedule, so they close the old CAPA and create a new one with a fresh timeline making it looks like actions are on time. These are completely unacceptable behaviors, create huge risks, and should be escalated.

It is important to have a tickler (a benefit to IT systems) to remind appropriate personnel to determine effectiveness at a predetermined time in the future. Failure to do effectiveness checks is a very common cause of Form 483 observations. Do not forget to do effectiveness checks!

Make sure improvements include controls to demonstrate continued effectiveness and sustain the gains. Organizations frequently have problems sustaining improvements. Memories are short. People change roles. The compliance pendulum swings. Improvements put in place are later removed by someone new in the job. And organizations go back to their old ways. The use of written control plans can reduce this common problem. Again, this goes back to the concept of process control. A well-managed process has a control plan that defines the key process input variables (KPIVs) and mechanisms for control. Keep that front and center when you make changes to processes over time.

Effective risk management is a prerequisite to effective and efficient improvement. Use risk management concepts to define priorities and assign resources. All nonconformities do not need to be treated the same. Actions must be commensurate with risks. Low-risk problems may be dealt with by correction only, as long as the basics of a QMS is in place. Medium-risk issues deserve the full rigor of a formal, closed loop CAPA. High-risk issues deserve additional project management, change management, and management review to ensure real, sustainable improvements. Set clear expectations for your organization.

Just like any other process, it is important to have metrics to monitor the performance of your improvement and CAPA process. Are CAPA activities timely, appropriate to the level of risk, monitored, and effective? Consider how many CAPAs are in the system, the type, status, progress to milestones, aging, and percent closed as effective. Put metrics in place to understand the health of your CAPA system. Your entire QMS depends on it. Your customers depend on it!

Red flags to watch out for in your CAPA and improvement processes are:

- Crisis management and firefighting are the normal modes of action.
- Priorities are constantly changing. People are constantly being reassigned or spread across a multitude of projects. "I spent 6 months working on that project and they just canceled it. What a waste, we were just making some real progress."
- Many low-risk projects managed in parallel with high-risk CAPAs derail the progress of high-risk CAPAs.
- Improvements are not finished but left hanging when priorities are changed, and people are moved to new projects.
- The same problems come back again over and over.
- Employees are frustrated by having to do things over and over again. "We never have time to do it right, but we always have time to do it over."
- There are little to no resources allocated to prevention, only to correction.
- Repeat audit observations are a big, waving, flapping red flag!

Capabilities for CAPA/Improvement can be characterized in the following table:

Learning	Succeeding	Thriving
CAPA system not fully established and maintained	CAPA system is at a basic level. It is established at a bare bones level. Many CAPAs are late, missing documentation, etc.	CAPA system used rigorously to reduce risks.
CAPA and improvement activities are focused only on correction and corrective action.	Lack of prioritization and effective management results many aging and ineffective CAPAs.	There is a shift from reaction to prevention. Focus shifts more and more to preventive action.
Correction and corrective action address only the identified nonconformity.	Further investigation is done to identify any other instances of the same or similar nonconformity.	Corrective action always includes remediation of historical or legacy documents.
Effectiveness checks not are not routinely done, have low standards of effectiveness, or are ineffective.	Improvement activities sometimes but inconsistently include remediation of historical or legacy documents.	Structured improvement process. Includes controls to sustain the gains.
		Change management is emphasized and used to enhance improvements.
There is no improvement process.		There is a common language of quality and improvement.
There are no root cause analysis or improvement tools.	Root cause analysis and improvement tools are used sporadically and inconsistently.	There is a well-equipped toolbox, lexicon of improvement, and well-trained resources for root cause analysis and improvement.
There is a void of sponsorship.		

In conclusion, remember the basics of **CAPA** and Improvement:

- If you did not fix it all, you will have to fix it all later.
- If you do not fix it right the first time, you are going to have to fix it again, and again, ... and AGAIN. That is the height of inefficiency.
- It is a lot more work to fix things after you have gotten a Form 483 or warning letter.
- Put controls in place to sustain the gains. Do not mess up the controls later on.

SHARE AND COMMUNICATE

S is for share and communicate. Good data and information is necessary for good decisions. Quality objectives need to be communicated throughout the organization and incorporated into functional objectives, process ownership, and down to individual objectives. Metrics need to be rolled up and communicated in management review. Risks need to be clearly articulated so that management can take actions commensurate with risk. Do not underestimate the importance of sharing information in an open and transparent manner. Yet, many companies struggle with this simple concept. And there are steep consequences because of it.

There are many places in the regulations that explicitly require sharing information:

- Management review requires sharing information with management regarding the suitability and effectiveness of the QMS.
- CAPA requires that information is disseminated to those directly responsible for assuring quality.
- CAPA also requires submitting relevant information for management review.
- Throughout the regulations we see requirements for "review and approve".
- There are specific timelines for communicating adverse events, medical device reports, and corrections and removals to regulators.

These requirements are put in place because the regulators want to emphasize the importance of sharing good information. Management review is one formal mechanism for sharing information on the health of the QMS. But, it is far from the only mechanism. Be conscious about communication.

Communication takes place everywhere, every day, and at every level. It is the most basic, common, and essential part of every business. Communication is the vital link between management, employees, customers, and stakeholders. Open lines of communication require thoughtful deliberation and execution. Communication is an essential ingredient for an efficient and effective QMS. It is critical that data is shared accurately, openly, and transparently in order establish and maintain a suitable QMS.

In many cases, management is surprised when they receive a Form 483 or a warning letter. They are surprised even though they have been getting appropriate information through management review. When management is surprised, there is something wrong with the communication process. Either the information and warning signs were not well articulated, or management failed to heed the warning signs. Remember the "sender—message—channel—receiver" model of communication we all learned in school. It deserves conscious consideration within the context of an effective and efficient QMS. A break-down in any part of the communication process disrupts the entire communication process.

What we've got here is a failure to communicate.

From the movie *Cool Hand Luke*

CONSIDER SENDERS OF INFORMATION

Anyone in the organization can and should be a sender of information. Any manufacturing operator, engineer, administrative assistant, or other person can and should be able to send information about issues. A healthy QMS needs individuals that feel safe in sharing information. This communication needs to go both ways.

Senders often feel like they have been telling management over and over again about problems. They are frustrated by lack of response or suitable resources. Before the space shuttle Challenger disaster, employees raised concerns about the O-rings and weather conditions. One employee wrote a memo titled "Help!" In a meeting, another employee said he was "appalled, just appalled" by the decision to launch. But, for various reasons, these concerns were not addressed. Seven astronauts were killed, and the shuttle program was put on hold for 32 months. It is critical that quality and compliance leaders stay watchful for these kinds of signs and take prompt action.

When individuals feel unsafe or feel like they are unheard, they may also seek alternative roads eventually leading to whistle-blower calls to regulatory authorities. Employees of medical device companies have been known to call the FDA about serious issues, triggering a for-cause inspection. In order to reduce this risk:

1. Your annual quality survey should include some measure of how safe personnel feel about sharing information, identifying problems, and raising concerns.
2. Consider both formal and informal mechanisms to gather information such as a suggestion box or employee roundtable meetings.
3. Recognize and reward individuals that have made positive or innovative improvements. Consider creating formal awards programs for significant achievements.
4. Make it safe to escalate concerns.
5. Have an internal hotline for quality and compliance issues.

Other senders of information can be functional managers and process owners via process performance metrics, analytics, and reports. Functional managers and process owners are responsible for ensuring all information is complete and accurate. Make sure that they fully understand this responsibility. Provide training for all functional managers and process owners to prevent a breakdown in communication. Make sure that functional managers and process owners understand expectations for sharing complete, accurate, and statistically valid data and information.

Other examples of sending information include:

- Auditors must document observations in an objective and factual manner.
- Personnel preparing reports must ensure completeness and accuracy.
- Trainers must be skilled in sharing information. Check for training effectiveness.

- Analysts must ensure that data are prepared and structured in a meaningful way. There is an art to this and good analysts are skilled with visualization techniques to make information more clear and meaningful.

Of course, the customer is always a critical source of information. It is vital to have open, clear communication channels for your customers. ISO 13485:2016 has specific requirements for customer feedback in clause 8.2.1. Make it easy for your customers to share information and provide feedback. Provide multiple channels for customer input.

TIP

If you have a website "contact us" form, make sure that it is monitored for communications that are alleged complaints. These must be shared with the complaint handling unit in a timely manner.

Communications with regulators may be required in terms of reporting adverse events, regulatory submissions, and also during inspections.

Good practice is for senders of information to check to see if information has been received. If not, encourage people to change the channel or medium for the message. Encourage people to escalate the message. Remember, the known problem is always better than the unknown problem with respect to quality and compliance.

The single biggest problem in communication is the illusion that is has taken place.

George Bernard Shaw

Warning signs that your senders of information have not been heard, include the emails and language described in the Challenger example. Of course, you must be aware of exaggeration and hyperbole. That does happen, and organizations should be trained on safe writing techniques. But, do not rush to dismiss employee concerns! When big quality issues and compliance issues occur, there are often warning signs that were ignored. Be very cautious.

Also, be aware of warning signs that people are afraid of raising issues. People frequently attempt to minimize the frequency or severity of an issue. They massage the data so that it looks more favorable. Or personnel present the data in misleading ways that alter the ability to effectively interpret and understand the true risks. They fear getting poor performance reviews or small raises. Warning signs include comments like, "Management doesn't need to worry about the details."

CASE STUDY

One business site was performing below average in many key performance metrics. When compared to other entities metric by metric, yes, they looked like there was room for improvement. But each area, on its own, did not present a significant risk. Management was unwilling to address

(Continued)

CASE STUDY (CONTINUED)

issues. It was necessary to force full clarity of the data. The format of the information was changed to show all the metrics for this entity on one profile view. By seeing that this business entity had unacceptable performance in many areas, it was easier to highlight overall risk. Looking at multiple factors together creates a profile that makes it easier to see how risks add up.

Whether accidentally or deliberately done, be very watchful for misleading graphs and data with:

- Flawed correlations
- Deleted outlier points
- Missing intermediate time points
- Altered or incorrect scales on graphs
- Truncated data
- Diluted data
- Omission of records
- Misdirection.

Receivers of information: Management must take responsibility for actively listening to and receiving appropriate information and data. This means making an intentional and focused effort to understand information. Again, this a key concept of management responsibility and strict liability to actively create and maintain a suitable QMS. Management can facilitate a data driven culture by asking the right questions:

- What is the issue?
- What is the risk?
- What does the data say?
- Do you have more detail?
- What trends are you seeing?
- Is that change significant?
- Why is that happening? Why? Why? Why? Why?
- How will you improve it?
- What are your plans to address it?

Management must pay attention and ask questions to determine the suitability and effectiveness of the QMS. Do not dismiss voices of dissent. Pay attention to non-verbal cues as well. Management must heed the warning signs! And there are always warning signs.

The most important thing in communication is hearing what isn't said.

Peter Drucker

Management review is an explicit and essential requirement of the regulations. 21 CFR 820.20 (c) requires that "management with executive responsibility shall review the suitability and effectiveness of the quality system at defined intervals and with sufficient frequency according to established procedures to ensure that the quality system satisfies the requirements of this part and the manufacturer's established quality policy and objectives."

Like everything else in the QMS, the results of management review must be documented. The documentation must include meeting minutes, a summary of action items, and a statement about the suitability and effectiveness of the QMS.

Regular and rigorous management review is not only a regulatory requirement, it is an extremely important method for management to be informed and engaged in the health of the QMS. Allow adequate time for this very important task. Do not relegate management review to an hour at the end of your annual planning session. Allow time for questions and discussion. Management review, like any other process, should have documented procedures and performance metrics. Depending on your organizational structure, management review may be conducted at several levels such as plant, business, and corporate. Your procedure should define the levels and the required management representation. An effective management review process should include escalation criteria to determine what information needs to be escalated from one level to another. This should be based on the risk and required input from higher level management.

Frequency of management review is not prescribed in the regulations. Most companies do management review from one to four times annually. Quarterly seems to be the norm and is a reasonable period for gathering data, taking action, and having results to report.

A typical management review might include:

- Review of previous action items
- Items escalated form lower level management review
- Quality objectives
- Quality metrics and dashboards
  - Process performance
  - Product performance
  - QMS performance
- Quality audit results
  - Process performance
  - Critical observations
  - Repeat observations
- CAPAs
  - Process performance
  - Critical CAPAs
  - CAPAs that were ineffective
- Training and resources
- Complaints and medical device reports
- Post market surveillance

- Recalls and field corrective actions
  - Status
  - Progress
  - CAPA
- Customer feedback
- Indicators of customer satisfaction
- Supplier performance
- Results of external inspections
  - Summary data
  - Response progress
  - CAPA status
- Regulatory update
- Resource requirements and updates
- Changes impacting the QMS
- Other issues, liability, off-label use, key customer issues, etc.
- Improvement projects and status
- Summary of risks and need for escalation
- Conclusions
- Quality policy review
- Confirmation of the suitability and effectiveness of the QMS

In order to maximize effectiveness, ensure that a quorum is present for management review. Take meeting minutes and assign action items with due dates. Follow-up on action items at the next review. Ensure sufficient detail and adequate time for meaningful discussion.

> **TIP**
>
> Make sure that personnel preparing management review information are skilled in statistical techniques and present information in a valid and accurate manner. Use of visualization tools such as color coding, mapping, trend indicators, and more to enhance understanding of information.

Consider all communication channels essential for highly effective and efficient quality and compliance management. Create communication plans for essential quality and compliance information. A communication plan for essential data will include the steward of the data, the content, medium, frequency, receivers, and if necessary, escalation of information.

For example, the CAPA system should have defined sources of data that are analyzed and transmitted to the CAPA system. Each source of data should have a communication plan that describes the details of the necessary communication, including escalation. Required information (e.g., complaint records or batch history records) can be controlled via transactional data bases that can help error-proof required information. They can send notifications, approval requests, and reminders per your communication plan.

Consider the content of communications. Make sure that information is clear and data driven. Information shared (especially in management review) must always be fact-based, complete, and accurate. Any attempts to hide, delay, deny, dismiss, deflect, obfuscate, or otherwise massage information must be taken seriously and escalated.

Consider communication of changes. Careful communication is essential for controlled implementation of changes. Clear, defined expectations for communications result in successful and consistent change control.

As you evaluate your capabilities with respect to communication, consider:

- Is information shared openly and transparently?
- Is management informed, actively engaged, knowledgeable?
- Are there warning signs that people are afraid to share information?
- Are there warning signs that people feel unheard?
- Does management respond with appropriate understanding of risk level?
- Does management provide appropriate resources?
- Does management understand the value of quality?
- Does management know KPIs?

Communication is the most basic of human skills. Use it deliberately and consciously within your QMS. It is a prerequisite for a highly effective and efficient QMS. Create communication plans for key processes, metrics, etc. Using your risk management tools, determine escalation criteria for sharing high-risk issues. Proactively assess and improve communications using an annual culture of quality survey. Create channels for sharing lessons learned and opportunities for improvement. Companies that are world class must have multiple methods for ensuring good communication.

Red flags to be watchful for:

- Listen for signs that people feel unheard. Warning signs include comments such as, "We've been telling them for months, but they just won't listen" or "we knew about it the whole time."
- Watch for attempts to keep information out of management review.
- Look for attempts to hide, dismiss, minimize, dilute, deflect, delay, misdirect, or obfuscate issues.
- Deliberate attempts to falsify data are a big, waving, flapping red flag.

Capabilities for Share and Communicate can be characterized in the following table:

Learning	Succeeding	Thriving
Personnel do not recognize appropriate communications.	Quality and compliance data are shared usually with openness and transparency. There may be rare instances of	Quality and compliance data are shared with an expectation of candor and straight talk.

(Continued)

Continued

Learning	Succeeding	Thriving
	messaging, minimizing, deflecting, or obfuscating unfavorable information to make it look more favorable.	
	Attempts at messaging, minimizing, deflecting, diluting, obfuscating unfavorable information are frowned upon, escalated, and used as learning opportunities.	Forums exist for sharing best practices.
Emails exist that are titled with "Help," "Disaster," or other indicators of frustration and unsafe writing.	Annual culture of quality survey identifies and addresses communication issues.	Forums exist for sharing lessons learned. Forums exist to interact with leaders and ask questions about quality and compliance.
Personnel are afraid to raise concerns or identify problems.	Communication plans with method, criteria, and escalation method are implemented for all sources of quality and compliance data.	Mechanisms exist for formal escalation of concerns.
Communications are limited to regulatory requirements such as management review. Quality and compliance data is not consistently shared and reviewed.		Management clearly and regularly promotes a data-driven culture with openness and sharing of information. Newsletters, town halls, and lunch-and-learn techniques are used to encourage openness and transparency.
Customer feedback is limited to complaints.	Some consideration is given to end-customer input when designing new products.	Feedback from customers is proactively sought. The concept of the big "Q" broadens and enhances the value of customer feedback.
Management Review is conducted superficially and seen as a regulatory burden.	Management Review is led by the Management Representative. The Management Representative provides data to evaluate the effectiveness and suitability of the quality	Management Review is guided by the Management Representatives. Functional leaders and process owners provide clear, complete, and transparent data to

(Continued)

Continued

Learning	Succeeding	Thriving
	management system. Functional managers or process owners attend but may be defensive.	evaluate the effectiveness and suitability of the quality management system.

In conclusion, remember the basics of Share and communicate:

- Every employee needs to feel free to speak up about quality and compliance.
- Process owners and functional leaders need to share information with openness, transparency, and candor! Attempts to minimize the magnitude or severity of issues will back-fire in the long run.
- Management that does not actively seek and listen to internal signals will eventually have to listen to Form 483 observations.

FUTURE TRENDS FOR SHARE AND COMMUNICATE

The Internet of things and modern EQMS systems have the ability to enhance data transparency, speed sharing of information, and improve escalation of risk. New technologies can improve collaboration, clarity and accuracy of data, and enable a culture of effective communication. They can automate sharing of information based on risk, escalate late action items, and enhance real-time sharing of quality and compliance performance data.

Tools to inform management review will improve in the future. They will facilitate real time metrics and dashboards that can be used in management review. They will allow easy analysis of data, segmentation, rolling up, trending, etc. real time in response to management questions. Companies that have leading edge capabilities in the future will mine (pun intended) these opportunities.

MEDICS SUMMARY

In conclusion, the **MEDICS** are essential for enabling a healthy QMS.

- **M**onitor—Ability to measure, monitor, and analyze the health of the QMS.
- **E**mbrace—Ability of companies to embrace a culture of quality, compliance, and prevention.
- **D**efine—Ability to define risks, prioritize issues, and drive key improvement activities.
- **I**dentify—Ability to self-identify problems and nonconformities with a short feedback loop.

- **CAPA and Improvement**—Ability to fix problems robustly and establish controls to sustain the gains.
- **Share and Communicate**—Ability to share and communicate key information in a transparent manner, especially management review.

These six capabilities are all essential for an efficient and effective QMS. Problems with any of these critical capabilities will reduce the efficiency and effectiveness of your QMS. Companies can reach significantly higher performance of the QMS by evaluating their individual MEDICS capabilities and developing improvement plans.

Always use thorough root cause analysis techniques like "5 Whys" to understand contributing causes of serious quality and compliance issues. Immaturity in the MEDICS is common contributing causes.

Quality leadership and a seat at the table

IV

Compliance must result in improved quality

Medical devices are becoming ever more complex and market forces are changing. Speed to market and low cost have been rewarded by the market in the past, but quality performance is becoming a stronger influence. This is driven by:

- There is an increasing likelihood and severity of quality failure due to the increased complexity of devices and use environments.
- There is increased visibility and cost of quality failure due to media focus, social media sharing, and negative publicity resulting in loss of reputation and sales.
- There is an increasing transparency of information and a drive toward comparative quality information.

In today's environment, these forces make quality and compliance ever more important. In the highly regulated medical device industry, it is important to have a clear understanding of the regulations. The quality and compliance organization can play a key role in interpreting the regulations to make sure that they are appropriately translated into company policies and procedures. In many cases, the regulations are brief and not entirely clear, requiring interpretation and translation. Through monitoring of the external environment, the quality organization can provide interpretation of new issues and their risk impact:

- It is important to understand the impact and risks of changing regulations such as ISO 13485:2016 or new programs such as MDSAP (Medical Device Single Audit Program).
- It is important to be aware of new guidance documents such as "Public Notification of Emerging Postmarket Medical Device Signals," issued in December 2016.
- Monitor Form 483s and warning letters for other medical device companies or suppliers that reflect a new focus area for regulators.
- Monitor recalls and supplier issues for other companies to understand implications. For example, in 2011 prepackaged alcohol prep-pads, swabs, and swab sticks were recalled by Triad Group after it was discovered that they might be contaminated. The swabs were sold under a variety of brand names and included with several injectable medications and medical devices. This caused multiple recalls of devices.

Medical Device Quality Management Systems. DOI: https://doi.org/10.1016/B978-0-12-814221-9.00009-9

It is important to investigate the current quality issues to interpret and understand the implications of changing regulations. As discussed in Chapter 1, Regulatory Requirements, the evolution of regulation depends on specific issues. A wise quality and compliance leader realizes that monitoring the external environment is an important role for the organization to reduce overall risk.

Throughout this book, we have covered the regulations and their interpretation. Building and developing the necessary capabilities, and methods needed to successfully implement them are a function of the quality and compliance organization. Ultimately, these are the key benefits that the quality and compliance organization brings to the medical device company.

All of this is focused on creating an efficient and effective quality management system to ensure that customers receive safe and effective products. Compliance alone is not enough. Compliance must result in superior product quality and customer satisfaction. That is not only good for the customer but also for the company. Improved quality helps companies become more productive; receive fewer complaints; manage fewer investigations and corrective and preventive actions; and have lower quality- and compliance-related costs.

In 2011, the Food and Drug Administration (FDA) published the paper *Understanding Barriers to Medical Device Quality*. The FDA published the paper after review of device quality data and feedback from industry and other stakeholders. Quality and compliance leaders can benefit by reading this paper in its entirety. The FDA Case for Quality consists of three core components:

- Focus on quality considers compliance as a baseline expectation by including critical to quality characteristics that result in higher quality outcomes.
- Enhance data transparency aims to leverage various sources of data and makes it available for searching and analysis by external stakeholders.
- Stakeholder engagement seeks to collaborate with stakeholders and launch initiatives that vary from traditional regulatory oversight.

Quality leaders can help the medical device company to understand this rapidly changing environment and the need for superior product quality to earn customer trust.

Shifting from cost of quality to the value proposition

10

American quality control expert, Dr. Armand Feigenbaum, first described quality-related costs in the 1950s in a Harvard Business Review article. Dr. Joseph Juran further shaped the traditional view of the "cost of quality" using the concepts of poor quality and investment in good quality. Dr. Juran said it was essential to measure quality in dollars, the language best understood by upper management.

The cost of quality is the expense of doing things wrong.

Philip B. Crosby

Philip B. Crosby later argued that nonconformances are a significant driver of total cost. Crosby popularized the terms "cost of poor quality" (COPQ) or "cost of nonquality," arguing that these costs could be avoided by up-front prevention. In *Quality Without Tears*, Crosby said that "The system for causing quality is prevention."

The cost of quality model breaks down costs into categories. It provides valuable thinking about costs associated with:

- Internal failure
- External failure
- Appraisal
- Prevention.

The costs of internal failure are costs associated with nonconforming product before it is distributed to customers. It includes dealing with nonconforming material, disposition, rework, scrap, and reinspection. It could also include costs associated with inefficient processes and activities that are considered "nonvalue added" by customers.

The costs of external failure are costs associated with product in the field that failed to meet customer requirements. It includes call center costs, complaint management, returned goods processing, adverse event reporting, field servicing, recalls, and potential liability management.

Appraisal costs are the costs of activities necessary to ensure conformance of products and services. This includes receiving inspection, in-process inspection, finished goods inspection, calibration, etc.

Prevention costs are those incurred to keep failure and appraisal costs to a minimum. These include quality planning, process controls, validation, training,

Medical Device Quality Management Systems. DOI: https://doi.org/10.1016/B978-0-12-814221-9.00010-5

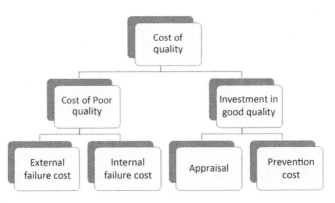

FIGURE 10.1

Traditional cost of quality model.

and internal audit. Because prevention reduces the costs of internal and external failure, it should be considered an investment. Training costs are usually a significant budget item and are always first to be cut when cost reductions are necessary. Training should be seen not as a cost, but as a strategic long-term investment.

Fig. 10.1 demonstrates the traditional cost of quality model.

These are good perspectives and it is useful for quality management to understand how their budget is allocated among these categories. Many companies find it easy to measure the COPQ but harder to measure the costs of prevention. Quality management should routinely understand where their resources in money and people are going. Think not only about where they are going today, but the optimum allocation for future success.

Additionally, the relationship is important, as investment in prevention and appraisal can drive down the COPQ.

EXAMPLE

Every recall has huge COPQs associated with it. The medical device manufacturer will bear the cost of failure investigation, and product scrap and disposal costs. There will be administrative costs for managing the recall activities, interacting with regulators, press releases, contacting customers, and exercising due diligence to retrieve all recalled product. There may loss of market share and a decrease in brand value. Perhaps the manufacturing line must be shut down for an extended period of time. A Food and Drug Administration (FDA) for-cause inspection might be triggered with a resulting Form 483 or warning letter. These costs can run into tens or hundreds of millions of dollars. The costs of prevention always pale next to these enormous costs. When seen with hindsight, a small investment in appraisal and prevention always seems like a good idea.

And there may be potential litigation costs as well. Huge payouts for class action suits are common such as those for pelvic mesh, bone cement, and even talcum powder. These issues can cost billions of dollars, make headlines, and are hot topics on news shows.

Adding COPQ to investment in good quality gives us the total cost of quality. In many industries, there is an attempt to find an optimal total cost of quality. This requires reducing prevention and appraisal costs to achieve a tolerable level of COPQ. However, a tolerable level of COPQ is not realistic for highly regulated medical products that have a critical impact on human health and safety. For medical products manufacturers must always strive to minimize external failure, no matter the effect on total cost of quality. For medical device, this means we need to shift from just the traditional cost of quality model and consider the value proposition for the products we make. A value proposition model includes not only the costs but also the benefits of quality.

Now we must consider what value quality adds in terms of customer satisfaction, improved sales, and market share. A long-standing company reputation of quality, compliance, and ethics enhances the company's brand value and trust mark. Consistent quality and compliance can decrease regulatory burden to the organization. In this way, the quality organization and the quality management system (QMS) can be seen as having a positive impact, or adding value, to the business.

In 2016, MDIC (Medical Device Innovation Consortium) released for review a whitepaper on the value of quality that included the benefits of quality. The benefit of good quality included cost savings and increased revenue. The paper was focused on C-suite and individuals who could use the concepts as a core management tool.

Fig. 10.2 shows a more appropriate model for the medical device industry. This can be used by quality leaders and management representatives to more clearly articulate an investment in prevention.

FIGURE 10.2

Value proposition quality model.

Quality leaders need to understand the cost of poor quality but increase awareness of the value proposition. Customer perceptions of quality can lead to increased trust, sales, market share. Products that are perceived as higher quality can command a higher price. Benefits of superior product quality include:

- Reduced COPQ
- Improved customer satisfaction
- Improved customer loyalty
- Increased market share and sales
- Increased customer trust essential to brand loyalty.

The best way to predict the future is to create it.

Peter Drucker

These benefits are difficult to measure and prove causality, but we will explore some approaches. Most industries have independent organizations that provide comparative product data. JD Powers has data on quality and reliability of cars. Consumer reports has information on a multitude of products from appliances to vacuum cleaners. The CNET website has reviews on consumer electronics. Better Business Bureau has information on business performance. There is a wealth of information on most consumer products.

But, medical devices are the exception. There is relatively little information on comparative quality. For many device types, consumers, health care providers, and payors do not have an independent or reliable source of device quality. The best sources of information are the FDA database for recalls and warning letters, medical device reports, and MAUDE (*Manufacturer and User Device Facility Experience*) database. But, these sources are incomplete. They only provide information on problems and little performance or reliability data. There is little data available on:

- Usage and failure rates
- Root causes
- Industry-wide patient outcomes
- Product outcomes
- Failure rates or customer satisfaction by product type or brand.

As part of the Case for Quality Program, the FDA is partnering with MDIC to pilot approaches for comparative quality information. Watch this space for future developments by attending various industry conferences or webinars.

Quality and compliance can also be seen as a key competitive differentiator. A reputation of quality and integrity builds brand loyalty and allows companies to command a premium for that. It takes a lifetime to build a reputation of trust and integrity. And it takes only one class I recall, warning letter, or evening news story to destroy that.

Use COPQ data to:

- Identify and prioritize improvement projects
- Trend sites, functions, business units for comparison
- Justify resources for improvements
- Inform management review
- Use as an additional data point to help employees understand the relevance and importance of their jobs
- Value of quality concepts are valuable tools for management review and for communication to all employees.

In practical terms, it can be difficult to measure the true value of quality and compliance. However, the concepts of the value of quality and compliance can be used to articulate a case for change and improvement. Understanding the magnitude and financial impact of a quality problem can help to justify resources. Over time, categorize costs, cost and prevention trends, and compare to market trends and customer survey results. Build a history of categories of cost over time. Showing a correlation over time between investments in prevention and lowered COPQ can be a very powerful tool for quality leaders.

Table 10.1 summarizes the costs and value of quality. Use it as you consider your current situation and your strategy for success. Assign your personnel costs (salary and benefits) to the categories below (e.g., how much is going to audit). Determine where your resources are going and where you want them to be in the future. Customize it for your organizational complexity and product risk levels.

In conclusion, cost of quality modeling translates quality into the language of business, money. The traditional cost of quality model needs to be modified to better fit the medical device industry. Adding the value proposition makes the model more useful. The costs of prevention may be more appropriately considered as investments in quality, driving down total cost of quality. Training, and

Table 10.1 Costs and Value of Quality and Compliance

Internal failure costs	Nonconforming material management
	Scrap
	Rework
	Repair
	Sorting
	Reinspection, retest
	Failure investigation costs
	Costs of inefficiencies
	Correction, corrective action, remediation costs
	Redesign
	Retrieving, correcting, remediating lost or missing data, documents, or records
	Nonvalue-added work

(Continued)

Table 10.1 Costs and Value of Quality and Compliance *Continued*

External failure costs	Returned products management
	Call centers and complaint handling
	Medical device reporting
	Field servicing cost
	Warranty repair
	Post market surveillance
	Recalls, field corrective actions
	Liability
	Hidden costs of lost sales
	Loss of reputation and market share from chronic poor quality
	Form 483 response
	Warning letter response
	Consent decree settlement costs, penalties, legal costs, third party inspection
Appraisal costs	Receiving inspection
	In-process inspection
	Finished goods inspection
	Supplies for testing and inspection
	Inspection, measuring, and test equipment, calibration
	Internal audit
Prevention costs	Process control and monitoring
	Design control
	Production and process controls
	Statistical process control
	Verification and validation
	Monitoring sources of data
	QMS development
	Standard work
	Risk management
	Quality planning
	Management review
	Training
	Document control
	Purchasing controls, supplier evaluation
Value of good quality	Reduced complaints
	Reduced COPQ
	Reduced regulatory inspection risk
	Reduced regulatory enforcement risk
	Increased customer satisfaction
	Increased customer loyalty
	Increased market share
	Increased revenue

other prevention costs, should be seen as a strategic investment in quality. By investing in quality, companies can reduce the risk of the staggering costs of recalls and regulatory enforcement/warning letters. An investment in prevention pays for itself in terms of improving BOTH effectiveness and efficiency of the quality management system. And that is good for the business!

Maturity modeling in medical device companies

11

Maturity modeling is an excellent technique to define current state and develop a future state vision and roadmap for change and improvement. Maturity models are simply a set of structured levels that describe how well processes can reliably and sustainably meet quality objectives.

The concept of maturity modeling has been around for a long time. In his 1980 book, *Quality Is Free*, Philip B. Crosby advocated the use of a Quality Management Maturity Grid (QMMG). This simple 5×6 grid showed 6 categories or aspects of quality management with 5 levels of maturity known as uncertainty, awakening, enlightenment, wisdom, and certainty. These levels were applied across the following six measurement categories:

1. Management understanding and attitude
2. Quality organization status
3. Problem handling
4. Cost of quality
5. Quality improvement actions
6. Summary of company quality posture.

It was used by asking a number of people in the organization to make a subjective judgment about which stage the company was at. Each check mark for uncertainty was given a score of 1, each checkmark for awakening was given a 2, and so on. The scores were added together for a total score. Today, there are a multitude of models with differing levels and categories.

From this old but useful concept, we can create more detailed maturity models for the processes and capabilities unique to the medical device company. Medical device companies can use the concept of maturity modeling and customize it for specific quality elements or processes. The use of maturity modeling helps to analyze and improve processes (e.g., the design control process or manufacturing processes).

There are many scales or levels of maturity that can be used. The scales may be applied to multiple processes and capabilities. As maturity levels increase, core activities become formalized, standardized, and result in improved outcomes.

Medical Device Quality Management Systems. DOI: https://doi.org/10.1016/B978-0-12-814221-9.00011-7

The higher the level of maturity, the lower the level of errors, nonconformities, or quality problems.

> **NOTE**
>
> In the MEDICS capabilities, I described capabilities using maturity modeling concepts. I used learning, surviving, and thriving as the three levels in the scale for each of the MEDICS. Possibilities for leading capabilities were described in the future trends section. For each of the MEDICS, I outlined characteristics for each level of maturity.

The MEDICS maturity levels can be used as the basis for further maturity modeling of your QMS processes, especially product realization processes. Use maturity modeling to describe the characteristics of maturity for various quality management system (QMS) processes. Evaluate your current state and target future state. From there, identify gaps and define the needed actions and plans to get to the targeted future state. These may include multiyear plans and should be included in your quality objectives.

The maturity levels in this chapter are examples only. You can create your own levels, definitions, and rating scales appropriate for your business. Some organizations use levels such as instability, stability, and maturity. Another example is novice, beginner, mature, advanced. More levels can be used for greater discrimination, especially for product realization processes. Capability Maturity Model Integration (CMMI) is a very commonly used model.

- CMMI levels:
  - Initial: The process is characterized as ad hoc, inconsistent, and occasionally even chaotic. Defined processes and standard practices, to the extent that they exist, are summarily abandoned during a crisis. Success depends on individual effort, talent, and heroics.
  - Repeatable: Processes are planned, documented, performed, and controlled at the project level.
  - Defined: Processes are well characterized and understood and are described in standards, tools, and procedures.
  - Quantitatively managed: Processes are controlled using statistical and other quantitative techniques.
  - Optimizing: There is a continual improving process performance through both incremental and innovative technological improvements.

In a June 2015 paper, *Maturity Model Research Report*, MDIC (Medical Device Innovation Consortium) considered the CMMI the best fit for medical device companies. See the future trends section regarding an FDA pilot using this model.

Maturity modeling has many benefits:

- Defines standards of excellence
- Creates aspirational goals
- Creates clear criteria for assessment
- Provides a basis for improvement projects
- Provides a basis for training and development programs
- Sets a foundation for reward and recognition.

The beauty of maturity modeling is that it demonstrates to the organization the standards of excellence. Functions and individuals have clear criteria for what success looks like. The maturity model itself provides a vision of what is possible. It can be a helpful mechanism to initiate change management in a positive way.

Maturity modeling provides the organization with aspirational goals. Maturity modeling and assessment gives teams, functions, and processes a goal for improvement. It allows process owners and functions to see possibilities without new rules being forced on them by the quality organization. Functions can see how they compare to best-in-class organizations; they can understand their own gaps and improvement opportunities. It can even create friendly competitiveness among sites/functions to reach higher levels.

Maturity modeling provides tools and training for improvement. By knowing the current maturity levels and gaps for a process, specific training can be developed or brought in-house. Training can be targeted for key gaps. A curriculum of training can be identified and phased in.

Maturity modeling recognizes and rewards improvement. By conducting maturity assessments, it is easy to see which entity being assessed (e.g., which new product development team) has improved the most. It is easy to see which individuals have achieved certain levels of subject matter expertise. Functional leaders/process owners can create awards and recognize results such as most improved, or bronze, silver, and gold levels. There are many ways to use maturity modeling to reward process improvement and excellence.

Maturity modeling and assessment is a key mechanism for driving standards of excellence throughout the organization and improving QMS efficiency and effectiveness. Processes become more consistent, reliable, and predictable. It is an invaluable method of preventing quality and compliance problems. And that results in a QMS that becomes more effective and more efficient.

Maturity modeling can be applied to any process including preproduction, production, and postproduction processes. Best practices are for each function or process owner to create their own maturity models for process excellence. For each of the processes defined in your quality manual, there can be a maturity model. For example, some companies have very detailed maturity assessments for design for quality programs. Other companies have very detailed assessments for manufacturing excellence or supplier excellence.

CASE STUDY

A maturity model for design reliability might be used to enhance the current waterfall model of the FDA with detailed assessment of:

- Design planning
- Design inputs
- Voice of the customer
- User needs and intended uses
- Design for manufacturability
- Design for reliability
- Understanding conditions for use
- Human factors
- Risk management
- Robust design analysis and methods
- Quality function deployment (QFD) or design thinking
- Design output
- Requirements flow-down and traceability
- Design capability
- Reliability planning
- Supplier evaluation and control
- Verification
- Validation
- Design transfer
- Design review
- Use of advanced tools
- Real-time assessment of documentation accuracy and completeness
- Collaboration and teamwork
- Training and engagement of team members

A medical device company can create assessment criteria and maturity levels for each of these categories. The assessment can be applied to various development projects/teams to understand the center and spread of capability. There may be a gap discovered, for example, with understanding conditions for use that result in design validation that is inadequate. Based on this information, additional training can be developed and provided, raising the overall skills of the organization. Perhaps one development project is lagging behind in one category. This could be managed by adding subject matter expertise or coaching to the team. Perhaps, another team has developed a best practice for managing traceability of requirements. A best practices lunch-and-learn session could be arranged.

There are different ways of assessing maturity. In some companies, self-assessment and auditing are used. In other companies, maturity modeling and assessment depends on independent appraisers to execute appraisals, provide feedback, and guide future activities. Other companies create functional councils that define their aspirational goals and program approach. These assessment programs are linked to training courses and development opportunities that provide tools and resources for improvement.

To use maturity modeling successfully, consider the following steps:

- A process owner or functional leader begins by creating a council for process XXX excellence (e.g., design for quality or manufacturing excellence) governance.

- Create a framework with key levels of maturity across categories for maturity assessment.
- Identify key process aspects for assessment.
- Define appraisal data and performance measures.
- Create a playbook for appraisal.
- Create courses, development opportunities, training, and coaching mechanisms to aide assessment and improvement.
- Use visualization tools (scorecards; heat maps; chicklet charts; red-, green-, and yellow-coded speedometers) to highlight strengths, opportunities, and improvements over time.
- Create incentives, recognition, and rewards for improvements.

Maturity modeling can also provide the basis of a systems approach to managing the elements of your QMS as a unified whole. It can be used to ensure that your system of objectives, processes, metrics, and improvement activities work in an integrated, unified manner. Maturity modeling (along with metrics) can help you assess your progress in individual quality system elements and as a whole. The Shingo model and Malcolm Baldrige quality awards are good examples to review.

In conclusion, maturity modeling is a powerful tool for defining levels of excellence and aspirational goals. It can be a powerful method of encouraging process/functional owners (such as R&D, manufacturing operations, or supplier management) to create their own programs for process excellence and maturity. A system of assessment, improvement plans, and training and development, combined with reward and recognition programs, encourage process ownership and excellence. And that is good for an effective and efficient QMS.

FUTURE TRENDS

As part of their *Case for Quality* (CfQ), the FDA has announced its intention to conduct a voluntary CfQ pilot program to explore the effectiveness of a quality maturity appraisal. The pilot was formally announced on December 28, 2017, in the Federal Register, Volume 82 FR61575, reference Docket ID FDA-2017-N-6778. The FDA, in collaboration with MDIC, has developed a maturity model and appraisal system that uses the CMMI system administered by the CMMI Institute. The CMMI Institute will administer the program. At the time of writing this book, the FDA was in the process of seeking volunteers and conducting the pilot. As an incentive, the FDA will forego conducting routine surveillance inspections of the volunteers. Detailed results of the assessment will be provided to the manufacturer for improvement. Summary results will be provided to the FDA. The program is named Medical Device Discovery Appraisal Program (MMDAP).

SUMMARY—A SEAT AT THE TABLE

Concepts of quality and compliance have developed significantly from early days of quality control and quality assurance. World-class quality and compliance requires that medical device companies have a system of intelligence to define, measure, and articulate quality and compliance risks and opportunities. It is more common now for quality, compliance, and regulatory leaders to have a seat at the table to bring this intelligence to the table. Business success depends on analyzing, interpreting, and identifying global risks and leading risk reduction activities. Executive decision makers need quality and compliance input to interpret risk profiles and provide intelligence to increase revenue and sales. Understanding regulatory threats, mapping risks, and defining go-to-market strategies are the values that the quality organization brings to the table. Creating an efficient and effective QMS that yields compliance, quality, and predictable results is the goal.

To earn a seat at the table, quality and compliance leaders should define their approach to:

- Facilitate interpretation and translation of regulations and constantly changing expectations.
- Identify and mitigate global quality and compliance risks.
- Help the organization transition from a focus on compliance to improved quality outcomes.
- Transition leadership from cost of quality thinking to investment in prevention thinking.
- Communicate the value proposition of quality.
- Use maturity modeling and assessment to define desired quality system and functional excellence characteristics and capabilities with aspirational levels of maturity.
- Build processes for training, development, and coaching to reach standards of excellence.
- Develop reward and recognition programs around standards of excellence.

Vision, strategy, quality planning

Compelling vision

You are probably already so busy doing things that you have little time to breathe, much less plan. But, plan you must. You must take a small portion of your time to think about the future and what it looks like. Think about your future and how you can realize what you want it to be.

> *You've got to think about big things while you're doing small things, so that all the small things go in the right direction.*

Alvin Toffler

A suitable and effective quality management system (QMS) takes good planning. We have already discussed the concepts of management responsibility and the roles of executive management and the management representative. Now, we must move on to how these roles can translate quality objectives to plan and establish an effective and efficient quality system. In Chapter 4, QMS Structure, we discussed the structure of the QMS; now we need to discuss how to create and implement such a structure.

As a reminder, the Quality System Regulation (QSR) defines the quality policy as "the overall intentions and direction of an organization with respect to quality, as established by management with executive responsibility." ISO 13485:2016 has slightly more detailed requirements in clause 5.3 Quality policy. "Top management shall ensure that the quality policy:

a. Is applicable to the purpose of the organization;
b. Includes a commitment to comply with requirements and to maintain the effectiveness of the quality management system;
c. Provides a framework for establishing and reviewing quality objectives;
d. Is communicated and understood within the organizations;
e. Is reviewed for continuing suitability."

So, your quality policy defines where you are now. It defines the commitment and objectives for the organization. But, it does not define where you are going. Of course, you will always have to do the things defined in the regulation above, but hopefully you want more than that. This is the point at which one needs to start thinking about a vision of the future.

It is important to identify the challenges and opportunities that the future brings. As stated in Chapter 2, Increasing Expectations, regulators continue to increase their expectations for medical device companies. Customers are becoming more

Medical Device Quality Management Systems. DOI: https://doi.org/10.1016/B978-0-12-814221-9.00012-9

educated and demand safe and effective products. Business stakeholders require that the QMS is not only effective but efficient. Quality and compliance leaders need to respond to these stressors to improve QMS efficiency and effectiveness. Thinking about the future starts with understanding the current status.

SCAN THE INTERNAL ENVIRONMENT

Start by analyzing the current internal environment. Be aware of the compliance pendulum! This is illustrated in Fig. 12.1. Many organizations shift wildly from a complete disregard for quality and compliance to a state of panic in the face of regulatory enforcement. When things are going well, management wants to focus on revenue generation activities and selling products. They put a strong emphasis on sales, marketing, and new product launches. Other activities are minimized. The organization becomes complacent about quality and compliance. The budget for quality and compliance slowly gets cut back. Corrective and Preventive Actions (CAPAs) remain open longer, audit observations creep up, and the organization slowly drifts away from a culture of quality and compliance. Risks accumulate and then management is surprised by a recall, a Form 483, or even a warning letter. It feels like a sudden occurrence, but the signs were there all along and ignored. Panic ensues. In that state of panic, there is a true emphasis on quality and compliance, but actions are not well coordinated or efficient. Risk mitigations and corrective and preventive actions all take resources away from revenue-generating activities. In the face of a Form 483, warning letter, or consent decree, additional resources, third-party auditors, consultants, and corporate experts are brought in. Extra reviews and approvals, redundancies, and nonvalue-added work are introduced into processes with little improvement in quality and compliance and sometimes at significant cost. Process and organizational misalignments may occur. When things finally stabilize, the QMS is seen as burdensome and restrictive. Management wants to focus on revenue generation. Lean projects are initiated to streamline the QMS. And the vicious cycle continues.

"We got a Warning Letter!!! Hire more people. You can have all the resources you want."

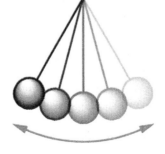

"We didn't get a Warning Letter. We can't afford to implement that CAPA/ redesign that product/ get that new IT tool for complaint handling/or xxxx."

FIGURE 12.1

The compliance pendulum.

A key role of the quality and compliance leader is to minimize the swings of the compliance pendulum. Neither side of the pendulum is efficient or effective. Stay in the sweet zone. Staying in the middle of the pendulum requires all of the quality system MEDICS:

- **M**onitor: Ability to accurately monitor and assess the current state of the QMS and keep it in the sweet zone, neither inefficient or too lean
- **E**mbrace: Ability to embrace and maintain a focus on quality while simultaneously meeting other business objectives
- **D**efine: Ability to define and articulate accumulating risks and prioritize issues
- **I**dentify: Ability to self-identify problems
- **C**APA and Improvement: Ability to fix problems robustly and forever instead of relying on superficial fixes;
- **S**hare and Communicate: Ability to share and communicate key information and risk in a transparent manner.

The successful quality and compliance leader depends on the ability to self-identify icebergs ahead, assess risk, prioritize actions, and fix problems in a lasting manner. Successful quality and compliance leaders need these capabilities to monitor, analyze, and communicate the health of the QMS in an open and transparent manner. They need to be able to accurately assess their current situation and obtain needed resources. The successful quality and compliance leader depends on a culture of quality and compliance, customer focus, and individual accountability to balance other business needs and challenges. Without these capabilities, the compliance pendulum can swing wildly.

It is important to recognize where you currently are in the compliance pendulum. You must accurately diagnose the situation. There are several different business situations that quality and compliance leaders typically contend with. Success depends on correctly diagnosing the situation. Each situation will present different challenges requiring a different strategy. Use the business situation diagnosis as an aid to evaluate your current state MEDICS maturity and needed improvement (see Table 12.1).

Table 12.1 Quality and Compliance Situations

Crisis mode	Organization is in a crisis due to a serious quality issue, recall, or significant regulatory enforcement actions.
Sustaining mode	Organization is performing well with acceptable quality and compliance results.
Too lean mode	Organization has insufficient resources and management commitment to maintain an effective management system.
Start-up mode	Organization is just initiating a quality management system and does not yet have product on the market.
Acquisition mode	Organization is being or has recently been acquired. Organization has unknown risks.

Organizations that are in crisis mode are easy to recognize. It is easy to see the state of panic that arises when a medical device manufacturer receives a significant Form 483 or a warning letter. This is even more so in the case of a recidivist warning letter, when the Food and Drug Administration (FDA) has required annual certification by a third party. And a consent decree will amplify the panic further. Significant enforcement activity can electrify management and send the board of directors into shock. The significance sinks in when management see words in a warning letter, such as:

- Failure to promptly correct these violations may result in regulatory action being initiated by the FDA without further notice.
- These actions include, but are not limited to seizure, injunction, and civil money penalties.
- Other federal agencies may be advised of the issuance of warning letters about devices, so they may take that into consideration when awarding contracts.
- Premarket approval applications for class III devices reasonably related to the firm's violations will not be approved until the violations have been corrected.

Chief executives may find the warning letter addressed to them personally. They may hear their own name and company issues on the evening news and see it in industry newsletters and notifications. They may be required to go to the FDA headquarters regularly to give updates on status of improvements. They may see their stock prices fall. When forced to sign a consent decree, they will be required to take actions, such as

- Reimburse the FDA for the costs of all inspections, investigations, supervision, reviews, and analyses the FDA deems necessary to evaluate defendant's compliance with this decree.
- Within 5 days of this decree, post a copy of this decree in common areas at company facilities and on the intranet website to ensure it will be reviewed by all company personnel.
- Within 10 days of this decree, deliver a copy to all directors, officers, agents, representatives, employees, attorneys, and successors with responsibility for the suitability of the QMS.

Of course, it can be an extremely painful and stressful time! However, the advantages of crisis mode are that it is easy to identify and it is easy to justify needed resources. However, it can be very difficult to provide direction and coordinate activities. Actions taken in crisis mode may not be well thought out resulting in lack of effectiveness, introduction of new problems, and creation of more opportunities for error. This creates a vicious cycle of problems and further enforcement.

Companies that are in sustaining mode have acceptable effectiveness in terms of quality and compliance results. They may or may not have acceptable efficiency results. They are in a state of conscious competence. But, they must be careful to monitor the suitability of the QMS to make sure they do not get too lean. In the absence of a crisis, executive management may attempt to streamline the QMS, tipping it to ineffective. Attempting to lean out an organization in sustaining mode must be done very carefully to avoid this risk. However, when companies recognize their healthy QMS, they have the opportunity to use quality and compliance as a competitive advantage. They can become a benchmark for quality and compliance excellence. They can use superior results to enhance trust with customers, increase brand value, and gain reputation as an ethical and well-managed company.

Companies that have QMSs that are too lean are in a dangerous situation. They are in a state of unconscious incompetence. They may think they are in a sustaining mode and not recognize their current situation until it is too late. They have developed a QMS that has some strengths and may have operated well in the past. But they have decreased resources too much. They may think they are performing acceptably, without seeing the dangers lurking below the surface. Risks increase slowly and may not be realized until it is too late. They are blindly unaware of their dangerous situation. Without awareness, they will be surprised by a Form 483, increases in complaints, customer dissatisfaction, or more serious quality issues such as a recall.

Start-up companies recognize that their QMSs are not yet suitable. They are in a state of conscious incompetence. They know that they have an incomplete and immature QMS. However, they may struggle with inadequate resources. They may prioritize other activities, such as technology or product development, over establishing a suitable QMS. If the QMS does not keep pace with other activities, the organization will take risks they do not know about. For example, they may base product design inputs and outputs on data taken from uncalibrated instruments, lack of laboratory controls, nonvalid statistical techniques, etc. They may not feel the impact of these risks for many years.

Acquisitions may be at any stage of quality and compliance. The challenge for the acquiring company is to accurately assess the situation and understand key risks. Many companies do due diligence to minimize the risks, but this provides somewhat limited information. They gain further information during integration activities, but by then it may be too late to get additional resources. If risks were not reported and needed resources not properly identified during due diligence, it becomes difficult to get resources later on. Executive management may feel frustration that the business proposition is now unacceptable if they must continue to invest more resources than originally planned. Or they may feel the acquisition target is lean and innovative and resist a perceived increase in bureaucracy. The company that was acquired may have their own resistance to change as well. There may be inadequate resources to integrate the acquisition in terms of information technology (IT) infrastructure. Do not underestimate the importance of integrating IT infrastructure. If IT infrastructure is not aligned, there will be long-term difficulties in terms of monitoring quality and compliance data resulting in unknown risks. It will be difficult to harmonize or standardize processes and performance. It will be difficult to share information and documents.

Table 12.2 summarizes the challenges and opportunities of these business situations. Each situation is unique, so use these characteristics as a starting point to evaluate key capabilities, focus areas, and develop a strategy for improvement.

Table 12.2 Challenges and Opportunities for Business Situations

	Opportunities	**Challenges**
Crisis mode	It is easy to recognize current state due to external quality issue or regulatory enforcement action.	Companies that are in this state may have inadequate capabilities for most or all of the MEDICS. Companies with recidivist issues or consent decrees may have executive management that is unaware, incapable, or uncaring of management responsibility.
	Continue to look for red flags (in addition to the ones that have already been communicated by regulatory enforcement).	
	After enforcement action, it is easy to justify resources for QMS improvement.	Resources may come all at once.
	Good planning and project management will optimize now plentiful resources.	Resources may come in the form of internal personnel temporarily reassigned from other areas (e.g., under a consent decree, companies may not get premarket approval. Unneeded personnel in product development are reassigned to areas needing remediation such as complaint review). Resources may come in the form of welcome consultants or unwelcome third-party auditors.
	Bring order to chaos.	
	Identify roles and responsibilities; needed skill sets and capabilities; and training plans.	Resources will have varying skills and capabilities.
		There is a risk of decreasing efficiency by adding redundancies, extra steps, extra approvals, etc.
		There is a risk of decreasing effectiveness further with hasty or poorly thought-out actions and CAPAs. Disconnected processes, lack of alignment, and nonvalue-added work create further opportunities for error.
	Management changes may be required for a turnaround.	It is difficult to change deeply engrained problems with management responsibility, lack of transparency and openness in sharing information, and unacceptable risk taking.
	Monitor the culture of quality and take deliberate steps to nurture and improve it.	Employees may be demoralized.
	Focus on improvement capabilities and restoring management responsibility.	

(Continued)

Table 12.2 Challenges and Opportunities for Business Situations *Continued*

	Opportunities	Challenges
Sustaining mode	Stay steady.	Business pressures may push the organization too far, swinging the pendulum again.
	Companies are in this state because they have adequate capabilities for all the MEDICS.	Do not become complacent and let capabilities deteriorate.
	Emphasize the value of quality. Careful analysis of results and resources is essential.	Resources may be more constrained as an organization attempts to optimize overall business success
	Develop leading-edge MEDICS capabilities to position the organization for growth. Use leading-edge capabilities to increase competitiveness.	Convince the organization that sustaining and improving is necessary and possible.
	Monitor future trends in industry. Use benchmarking to increase awareness of possibilities. Use brainstorming and visualization techniques to show a bright and successful future.	There is difficulty visualizing the possibilities.
	Focus on prevention to further enhance efficiency and effectiveness.	
Too lean mode	Assess and improve M (Monitor) and I (Improve) maturity to paint an accurate picture of risk. Look for red flags that risks have not been communicated to executive management with openness and transparency. Look for red flags of unacceptable risk taking. Accurate diagnosis is essential	The company may not recognize their current state. There is real danger of confusing this state with the sustaining mode. Companies are unaware of unidentified risks lurking below the surface. They are at risk of being surprised by a significant quality or compliance event.
	Improve self-identification, process monitoring, and improvement capabilities. Look for and address red flags of communication issues and lack of transparency and openness.	Companies in this state may have gotten there because they have inadequate capabilities in terms of self-identifying issues and providing adequate monitoring and measurements.
		- OR –
	Ensure management is aware of the concepts of management responsibility. Ensure management has the knowledge and skill sets to ask the right questions.	Companies in this state have acceptable M and I capabilities but have not effectively shared and communicated risk with executive management. Management is unaware of the true risks.

(Continued)

Table 12.2 Challenges and Opportunities for Business Situations *Continued*

	Opportunities	Challenges
		- OR - Management is aware of risk. Management has an unacceptable tolerance for risk and has failed to ensure adequate resources for a suitable QMS.
	Look for red flags of unacceptable risk tolerance.	
	Create a call to action based on risk analysis.	Resources are inadequate.
	There may be pockets of strength, skill, and capability that can be enhanced.	
	There may be acceptable processes still in place.	
	Focus on improved quality outcomes and reducing cost of poor quality!	
	Leverage teaching moments.	
Start-up mode	It is easy to recognize current state.	It is challenging to keep the QMS in step with business activities.
	People are energized by new opportunities and possibility.	Recognize what quality system elements are necessary for each business milestone.
	Focus on doing things right from the very beginning. Focus on preventing problems that will require remediation in the future. Remember, "If we don't have time to do it right, when will we have time to do it over?"	Make sure there are adequate plans and resources for an effective QMS on a step-by-step basis..
	There is an opportunity to build an efficient and effective management system without having to do correction, corrective action, and remediation.	It is challenging to build a QMS from scratch.
	Assess capabilities for all MEDICS. Look for red flags.	
Acquisition mode	Improve due diligence effectiveness. Create a due diligence checklist and process before it is needed.	Due diligence provides a limited assessment of risk.
	Expect that additional risks and required remediation will be identified during integration.	Integration plans and required resources are underestimated due to insufficient due diligence.
	Use change management and stakeholder alignment techniques	There is difficulty harmonizing and standardizing processes.

(Continued)

Table 12.2 Challenges and Opportunities for Business Situations *Continued*

	Opportunities	Challenges
		Insufficient planning for IT infrastructure (CAPA, complaints, document control, etc.) changes. Failure to bring acquisition to internal equivalence will create ongoing and long-term problems with consistency, accuracy, monitoring, and sharing of documents, data, information, etc.
	Use change management and stakeholder alignment.	Acquisition may be resistant to change. Or, they may be seeking change.
	Insist on the same monitoring and metrics for acquisitions to ensure apples to apples comparisons. Share and communicate key information, including integration status during management review.	Executive management may see the target as lean and agile. They will resist imposing a perceived QMS bureaucracy on the target, leaving them with an inadequate QMS.
	MEDICS capabilities are unknown. All MEDICS must be immediately assessed for a successful acquisition. Look for red flags.	The acquisition may need to grow a QMS that is efficient and effective.

Work to truly understand your company's history of quality and compliance. You cannot figure out where to take your organization if you do not know where it is and how it got there. Your vision will depend on where you start in the compliance pendulum. And it will depend on your current maturity and capabilities. It will also depend on your company's overall business vision and how quality and compliance support it. Use these inputs to determine your vision for quality and compliance. Your quality policy must enable your vision of the future.

SCAN THE EXTERNAL ENVIRONMENT

The next step is to scan the external environment. This is required to understand the regulatory climate, industry trends, market forces, supplier and competitor information. Understanding general trends and the world around you helps to determine your fit in that changing world. Information on these issues can be gathered from regulator websites, new sources, the Internet, industry groups, trade associations, customers, and suppliers. There are a multitude of industry conferences, seminars, and webinars on existing and emerging regulatory expectations. Scanning the environment requires ongoing effort. Important changes must be evaluated for impact on your strategy and plans. Some examples of general changes in the medical device industry, at the time of writing this book, are discussed below. Make sure you analyze your specific situation when you are ready to prepare your strategy and plans.

A huge focus in the United States today is on affordable health care. Medical device manufacturers have a role to play in this. Payers and healthcare providers are concerned about the cost and value of products. Patient safety and access to high-quality affordable medical care are key drivers for the healthcare system. Cost control, innovation, speed to market, and scientific evidence to facilitate reimbursement are key success factors.

Customers want better products, faster access to new technologies, and with acceptable cost. They want better information to compare treatment alternatives. They want to be able to compare product quality and outcomes. In most other industries, customers have access to comparative product quality. Consumers purchasing cars, cell phones, vacuum cleaners, and other products have easy access to product data from multiple sources such as consumer reports. But, consumers in the medical device industries have very little information. Doctors and healthcare providers may have a bit more knowledge and access to clinical data, FDA information, but they also have little information. Access to product performance data such as safety, effectiveness, usability, compatibility, and patient experience will become more important in the future.

The FDA has been actively promoting their Case for Quality program for several years now. FDA is conscious of costs and barriers to quality. FDA is moving from expectations of strict compliance only to sustained product quality. FDA is taking a collaborative approach with industry to enable and incentivize improvements. The FDA's Case for Quality program creates a compelling vision for change that includes the following:

- Conduct quality systems maturity appraisals to collect baseline effectiveness metrics.
- Use of product metrics may reduce inspection risk, remove manufacturers from the risk-based work plan, and waive preapproval inspections.
- Drive cycles of learning and improvement to increase QMS effectiveness. Enhance a continuous improvement mindset.
- Promote technologies to drive consistency and efficiency.
- Shift from records-based to data-based quality systems.
- Reduce the burden of validation.
- Medical device single audit program (MDSAP).

Medical Device Innovation Consortium (MDIC) is a nonprofit organization operating in partnership with the FDA to facilitate the FDA's Case for Quality initiative. Formed in 2012, MDIC "bring together representatives of the FDA, National Institutes of Health (NIH), Centers for Medicare and Medicaid Services (CMS), industry, and non-profits and patient organizations to improve the process for development, assessment, and review of new medical technologies." The

MDIC website provides additional details and valuable insights into new approaches and methods.

International Medical Device Regulators Forum (IMDRF) is a very useful source of information about new focus areas and changes such as MDSAP. Association for Advancement of Medical Instrumentation (AAMI), Advanced Medical Technology (Advamed), and the FDA website itself are all good sources of information about what is happening in the industry.

Scan the environment to understand how your competitors and suppliers are doing. If your competitor just had a major recall, what can you learn from it to improve your own QMS? Did they have a problem with a supplier that you also use?

We are in the fourth industrial revolution now. Technology, automation, data management, and integrated, electronic QMS systems are changing rapidly. The Internet of Things (IOT) and Industrial Internet of Things (IIOT) are common terms today. Consider how these new technologies and capabilities can improve your QMS efficiency and effectiveness:

- Technologies for supplier management, monitoring, and sharing information efficiently
- Technology automation to reduce errors and improve consistency
- Data management to enhance use of metrics, connect the dots, and understand the impact to the business
- Data management to facilitate real-time monitoring and control
- Technology for tracking and tracing products
- Product intelligence
- Change from documents focus to data focus.

Understand the impact of new technology on product risk and safety. For example, digital health is now a focus area for regulators. The FDA and IMDRF have issued guidance documents to clarify their position. The broad scope of digital health includes mobile health, wearable devices, telehealth, telemedicine, and personalized medicine. Providers want to use digital health to reduce inefficiencies and costs. Patients are hoping to better manage and track their health. Devices have the potential to connect and communicate with other devices. This potential offers many benefits. Risks such as cyber security need to be evaluated and mitigated. This space is advancing rapidly. Medical device manufactures need to stay on top of it.

Other regulatory changes include ISO 13485:2016 and the revised European Medical Device Regulation discussed in Chapter 1, Regulatory Requirements. Medical device regulations will continue to advance forward, based on changing product technologies, new issues, new risks, and social values.

Scan your specific market and customer needs. Identify trends that impact your QMS. Identify different product technologies (e.g., imbedded software or combination products) that will challenge your QMS.

ANALYZE STRENGTHS AND WEAKNESSES

Use the strengths, weaknesses, opportunities, and threats (SWOT) model (see Tables 12.3 and 12.4) to assess your current organization and quality system. This classic tool is well known because it is so simple and so useful. What changes need to be made to prepare your organization for the future? How do you stabilize the compliance pendulum to optimize the efficiency and effectiveness of your organization? Use the SWOT model to articulate your internal strengths and weaknesses and your external threats and opportunities.

Consider strengths and weaknesses based on the internal situation:

- What is your MEDICS maturity?
- How does your product quality compare to your competitor's?
- What is your organization's image and reputation for quality and compliance?
- What is your track record of quality and compliance?
- What is the condition of your assets?
  - QMS structure, processes, and procedures
  - Process capabilities
  - IT infrastructure
  - Employee skills and capabilities
  - Facilities, labs, technology, equipment.
- Understand your baseline quality dashboard.

Table 12.3 SWOT Analysis (Strengths, Weaknesses, Opportunities, and Threats) Model

STRENGTHS (focused internal situation)	WEAKNESSES (focused internal situation)
• MEDICS	
• Competencies	
• Compliance pendulum	
• Product realization	
• Product quality	
• Technology	
• Innovation	
• Efficiency	
• Effectiveness	
THREATS (focused on the external environment)	OPPORTUNITIES (focused on the external environment)
• Regulatory changes	
• Competition	
• Game changers	

Table 12.4 SWOT (Strengths, Weaknesses, Opportunities, Threats) Example

STRENGTHS (focused internal situation)	WEAKNESSES (focused internal situation)
• Sustaining mode	• Disparate sources of quality data
• Mature SOPs	• Inability to connect the dots in terms of data and impact of quality
• 0 recalls	
• 1 Form 483 with three observations	• D (Define) capabilities inadequate
• C (CAPA) and I (Improve) capabilities acceptable	
• Manufacturing technology	
THREATS (focused on the external environment)	OPPORTUNITIES (focused on the external environment)
• Case for Quality	• Move organization to leading M (Monitor), E (Embrace), and S (Share) capabilities to position organization for the Case for Quality
• Pressure to reduce cost of health care	
• Recruiting talent	• Evaluate MDSAP
	• Big Data
	• Integrated EQMS
	• Focus on prevention
	• Reduce cost of poor quality
	• Benchmarking to visualize leading-edge capabilities

Consider threats and opportunities focused on the external environment:

- New regulations or guidance
- Technology disruption
- Economic environment
- Market pressures

Future initiatives must be designed with the SWOT in mind. Your strategy must be designed to improve your weaknesses, reduce the risk of the threats, and use your strengths to capitalize on opportunities.

Do benchmarking to expand your view of possibilities. Explore potential technologies, solutions, successes and failures, and best practices. Learn from the success of others. Benchmarking is essential for visualizing and then improving leading-edge capabilities. Articulate best practices that can benefit your organization.

Your company may already have a business vision. There may be business objectives and challenges that impact the quality vision. Cost reductions and product innovation are common business goals. How does the quality system enable those goals?

Now it is time to develop the QMS vision. Imagine what the future looks like. Articulate the desired future. Create an aspirational goal.

If you always do what you've always done, you'll always get what you've always gotten.

Unknown

The QMS must enable the company vision. Some examples are as follows:

- Corporate and social responsibility depends on compliance with applicable laws and regulations.
- Compliance alone is not enough. Compliance must enable better product quality to accelerate growth, market share, or customer sentiment.
- Business success depends on a QMS that optimizes both efficiency and effectiveness resulting in overall business performance.
- Quality is an investment that increases customer satisfaction and trust.
- Quality increases brand value.
- Quality is an accelerator for growth.
- A reputation of compliance and ethical behavior is necessary for social acceptance and trust.

Creating a vision of the QMS in the future is best done as a team exercise. It needs to be integrated into your company's business vision. In fact, it must enhance or enable business success. Successful companies know where they are headed. Document your quality objectives.

The management representative should seek alignment with executive management regarding the vision. After all, management with executive responsibility is responsible for a suitable QMS.

In conclusion, we have used the concept of the compliance pendulum to define the various quality/business situations that a management representative might encounter. These business situations each represent unique challenges and opportunities. Use the characteristics of these business situations (see Table 12.2) to further assess your current state MEDICS capabilities and needed improvements to prepare for the future. Additionally, the various business situations represent different starting points for improved quality outcomes, improving the value proposition of quality and compliance, and improving QMS maturity. Apply the SWOT model to identify weaknesses, assess threats, and use strengths to capitalize on opportunities. Now, we can put these concepts together to create strategies and plans for improvement.

Translating QMS vision to strategic quality objectives and plans

13

Now that you know your starting point and have a vision of where you want to go, it is time to think about how you want to get there. There may be different paths. You can start thinking about the best path and what activities and projects are necessary. You can start thinking about the order of activities and the skill sets that are needed to get there. Strategy is about picking the best path and making decisions systematically based on situational analysis, environmental scan, gap assessments, and ongoing feedback.

The regulations put a lot of emphasis on planning. 21 CFR 820.20(d) requires that each manufacturer establish a quality plan which defines the "quality practices, resources, and activities relevant to devices that are designed and manufactured. The manufacturer shall establish how the requirements for quality will be met." This is a somewhat confusing requirement and companies have a hard time understanding if this means the entire quality system, the quality manual, individual product quality plans, or the company's annual planning and budgeting process. The Quality System Inspection Technique (QSIT) provides some guidance here. "Much of what is required to be in the plan may be found in the firm's quality system documentation, such as, the Quality Manual, Device Master Record(s), production procedures, etc. Therefore, the plan itself may be a roadmap of the firm's quality system. The plan in this case would need to include reference to applicable quality system documents and how those documents apply to the device(s) that is the subject of the plan."

Additionally, ISO 13485:2016 clause 5.4.1 Planning—Quality Objectives requires that "Top management shall ensure that quality objectives, including those needed to meet applicable regulatory requirements and requirements for product are established at relevant functions and levels within the organizations. The quality objectives shall be measurable and consistent with the quality policy."

In practical terms, quality planning encompasses many elements and takes place at many points. Use your quality policy, quality manual, and quality system structure and procedures to show your roadmap for how you plan for and establish a quality system relevant to the devices you make. Your quality manual

Medical Device Quality Management Systems. DOI: https://doi.org/10.1016/B978-0-12-814221-9.00013-0

183

should show your process structure and the relationship of processes. You can use the device master records (DMRs), design history files (DHFs), and risk management plans to show your quality plans for each product. Your system of metrics and dashboards, process monitoring, audit, and management review are used to assess suitability and plan for necessary improvement. Annual performance planning, communication of objectives, and objectives cascade to all individuals and are all part of the planning, alignment, and strategic improvement process.

Let us turn now to longer term actions to translate your vision into a concrete strategy and plans. Now is the time to develop your strategic or long-term quality objectives for how you want to achieve your vision of an effective and efficient quality management system (QMS). Match your strategy to your defined current situation and your identified strengths and weaknesses. The strategic quality objectives should articulate what you strive for and the means by which you will achieve it.

Systematically monitoring assets, making decisions, and defining actions require thoughtful and deliberate effort. It requires a road map of steps. Strategic planning should not be done in isolation. It is best done as a group effort. A skilled facilitator or consultant can greatly improve the process. Note: There are many approaches to strategic planning. It depends, again, on your organizational structure and approach. Your company may already have an approach. The approach explained below is intended to provide an example of how to apply it to a quality management for medical devices. You may choose to use an alternate approach.

When you have vision for quality in mind, think about the road map to get there. Use techniques such as a Merlin exercise to enhance the vision. Merlin was the legendary magician who helped King Arthur. In the book *The Once and Future King* by T.H. White, Merlin was described as living his life backwards. He could remember the future. Use this approach to remember what the future looks like and how you got there. Imagine yourself at the endpoint. Look back in time and remember what barriers you had to overcome to achieve that goal. Keep remembering what it was like when you were almost all the way there, and half way there, and a quarter of the way there. Use that to create the road map of how you got to your vision.

Another way of thinking of this is to imagine yourself at the bottom of a tree-covered mountain. It is very difficult to see the path to the top. But, when you are at the top of the mountain, it is much easier to see the starting place and the many paths that lead to the top. It is easier to see road blocks, steep paths, and barriers to success. From the view at the top, you can choose the best path.

The Merlin exercise is a great team exercise for visioning, planning, and alignment. Use it to describe the people, processes, technology, and MEDICS maturity necessary for each milestone. The power of the Merlin exercise comes from painting a vision of success and seeing solutions to road blocks that might occur.

The world that we have made as a result of the level of thinking we have done thus far creates problems that we cannot solve at the same level we created them.

Albert Einstein

Determine your strategy regarding MEDICS maturity. What level of maturity is necessary for each step on your journey? Consider how to improve the MEDICS individually and overall. Do you want to improve them one level at a time or leapfrog from learning all the way to thriving? How will you do it? Send staff out for training to give them the tools? Do benchmarking to help visualize the possibilities? Bring in consultants? Or bring in a new-hire change agent to shake up the organization. Maybe, dear reader, you *are* the change agent.

Determine your strategy to improve quality system *effectiveness*. For example, consider the following:

- A focus on prevention can be a strategy for improving both ineffectiveness and inefficiency.
- Product quality excellence:
  - Connect the dots (premarket, supplier, manufacturing, postmarket, etc.) for a complete picture of product performance.
  - Control points are established, with key performance indicators (KPIs), for each product.
  - Comprehensive risk management is in place for every product.
- Set maturity goals for value streams that drive superior product quality:
  - Supplier excellence and connection
  - Design for quality
  - Operational excellence.
- Develop training programs to address gaps in maturity.
- Implement reward and recognition programs for process excellence.
- Improve speed and responsiveness to customer issues.
- Shift priorities from reaction to prevention.
- Invest in IT systems that error-proof activities and actions.
- Focus on a full implementation of a process approach. Focus on process ownership and control. Every process has an owner. Every process has controls to ensure it is stable, capable, repeatable, and predictable.
- Increase QMS maturity and shift focus from finding to fixing to preventing issues.
- Communicate the value proposition for quality to optimize the value of quality.
- Develop the Big Q culture of quality.
- Determine plans for MEDICS improvement.
- Invest in training.

Determine your strategy to improve quality system efficiency. For example, consider the following:

- Promote standardization.
- Create centers of excellence.
- Reduce process variation.
- Shorten the feedback loop.
- Reduce redundancies, handoffs, and nonvalue-added work using value stream mapping.
- Implement error proofing, pokayoke, and mistake proofing concepts.
- Implement automation and technology solutions to improve efficiency.
- Integrate systems for improved speed, clarity, and transparency of information.
- A focus on prevention can be a strategy for improving both ineffectiveness and inefficiency!

Identify imperatives for business success. Consider actions that are necessary to enable next steps on the road map. For example:

- Gain management engagement.
- Conduct a culture of quality survey.
- Improve the culture of quality.
- Improve training systems.
- Develop an inspection preparedness program.
- Improve statistical techniques.
- Train on root cause analysis techniques.
- Obtain enabling technology to manage CAPAs (Corrective and Preventive Actions), complaints, calibration, etc.

These strategies should not be seen as "flavor of the month" projects. It is important to present them as part of a long-term strategy to reach the quality vision. As you roll out your strategy, be sure to highlight what results you are aiming for and how the initiatives and projects fit together. Define projects and improvements that are enabling or stepping stones for the next steps.

Spend time on identifying barriers to success. Are there things that are preventing you from moving forward? Perhaps it is difficult to get management with executive responsibility and other functional leadership engaged. In this case, it might be necessary to create a training session on management responsibility first. Bring in guest speakers who have experienced a quality crisis. Create opportunities for management to share lessons learned.

Are there prerequisite projects or activities that need to be completed first? Are there foundational processes that need to be in place and healthy before you can do other things? Identify and articulate the stepping stones. Create a road map or multiyear plan. Your road map should prioritize, and plan projects over time and optimize the use of your resources. Address how you will allocate your resources between sustaining activities (i.e., keep the lights on) and strategic

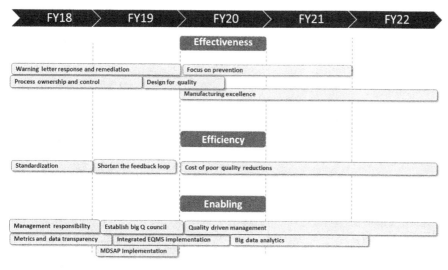

FIGURE 13.1

Example 1: Quality management system improvement road map.

projects. Fig. 13.1 below provides an example of a 5-year road map. In this example, there are projects for improving effectiveness, efficiency, and enabling ongoing business activities.

An alternative road map is displayed in Fig. 13.2. This example also provides a 5-year road map. But it breaks down improvement efforts into product quality, compliance, MEDICS, people, and tools and IT infrastructure.

It is vital to determine what resources (assets) you will need to follow the road map you envision. It can be useful to conduct a gap analysis. Do not assume that the assets you have now will be adequate to meet your vision. You may need to change your staff capabilities, your own capabilities, bring in consultants, etc.

Determine what resource gaps you have now and how that will change over time. Determine what skills and competencies will be required in the future. Identify the IT infrastructure needed now and for the future. Address needed equipment, facilities, tools, labs, etc.

WARNING

If you have a large organization, recognize that not all sites or units are at the same starting point. Because of this, they may have different needs and priorities. Failure to consider these differences will result in a significant lack of alignment. I have seen some organizations struggle endlessly because they have not recognized and addressed this problem. The gap analysis should articulate any gaps needed to bring less mature sites along. More mature sites may have to pave the way for overall business success.

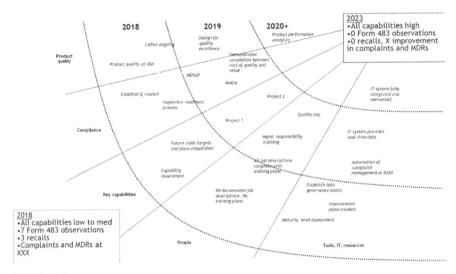

FIGURE 13.2

Example 2: Quality management system improvement road map.

RESOURCES

Planning for and obtaining adequate resources is an ongoing negotiation effort for scarce resources in your company. The preamble to the QSR (Quality System Regulation) comment 48 indicates that "resources emphasizes that all resource needs must be provided for, including monetary, supplies, etc., as well as personnel resources." Before beginning any negotiation for resources, the management representative must be in agreement with the management with executive responsibility on the business situation, vision, and quality plans. Never waste a good opportunity to justify resources. Both a warning letter response and a new product launch may require additional resources. When planning resources, consider people (headcount and skill sets), IT systems, travel, annual quality survey, training, consultants, supplies, computers, inspection preparedness plans, etc.

An assessment of MEDICS maturity, situational analysis, and a clear picture of risk are essential ingredients to ensure a suitable QMS. Resources needed will depend on your specific situation:

- In a crisis situation, resources are more likely to be available but must be carefully coordinated. Negotiate for authority to adjust or add resources as the crisis evolves. The full magnitude and extent of the crisis may not be readily apparent pending a full investigation.
- In a sustaining situation, you should already have clear information on your current state resources. Your organization may already be using an annual business planning/budgeting process with significant detail and historical

information. You must be prepared to justify keeping these resources. Most businesses face cost pressures that can potentially affect quality resources. Use value proposition for quality concepts to guide the negotiations. Use trend analysis to identify changes (e.g., new product launches or increases in complaints) to justify needed increases in resources. Use strategic planning to identify and justify incremental improvements.

- In a too-lean situation, you must carefully define a call for action for additional needed resources to improve your quality system and MEDICS maturity. A clear picture of risk and focus on improved quality outcomes will feature in your negotiation. Paint a clear and accurate picture of risk. Describe the known risks, the uncharted waters of the unknown risks, and inability to successfully mitigate known issues. Use management review to articulate the challenges and create a call for action. Use the call to action to justify resources.
- In acquisition situations, due diligence should highlight and articulate any issues that are show-stoppers, affect valuation of the deal, or need to be addressed during integration or long term. Pay close attention to needed IT infrastructure. Emphasize the high level of unknown risks relative to other business situations. The high level of unknown risks creates greater uncertainty in resources planning. Due to the high level of unknown risks, include a review at a specified future point in integration to review risks and renegotiate resources.
- A start-up situation requires a specific skill set with staggered resource additions at critical milestones. Relate needed skill sets and resources to key business milestones.

If you are reading this book because you are a quality or compliance leader in the organization, think not only about the resources in your own organization, but also about how resources in other organizations affect you. A huge cut-back in operations or R&D can create gaps in the quality system that you may eventually have to fill. You may end up having to plug the hole in the dike. If you are a functional leader or process owner reading this book, think about resources and assets needed to maintain process stability and control. Management with executive responsibility will need to make sure there is a balance of resources throughout the different functional groups. A key goal for resource planning is to optimize quality system effectiveness and efficiency.

If you don't have time to do it right, when will you have time to do it over?
John Wooden

Define needed investments in IT systems such as:

- Document control
- Production and process controls
- Calibration, maintenance, etc.
- Batch records or manufacturing execution systems

- Complaint management electronic reporting of medical device reports
- Nonconformance management
- CAPA management
- Manufacturing/operations
- Training
- Laboratory control
- Traceability
- Design control
- Requirements management.

Technology is an enabler of a quality system. New IT systems can bring significant efficiency and effectiveness improvements when done well. But, there is a significant investment to get them there. Clearly articulate the return on investment. Determine how you will implement and integrate IT solutions. Consider resources required for thorough validation and data migration. Poor implementation can also result in delays and unplanned consumption of resources.

Consider when and how these IT systems will be integrated. This is a rapidly changing area. There are automated total QMSs today that are very powerful and fully integrated. If you already have IT systems for separate processes, think about integration. If you are just getting started look for a fully integrated system.

There are always more needs than limited resources can deal with. Prioritization of projects is essential. The changes to ISO 13485 in 2016 recognize this realty and has multiple references to a risk-based approach. Actions must be commensurate with risk. Every company needs a consistent method to assess risk and assign priorities to quality objectives and desired improvements. The prioritization matrix (also known as a C&E (or cause and effect) matrix as described in Chapter 15, Alphabet Soup) is a useful tool for doing this. Use a prioritization matrix to analyze your potential improvement projects. This is an excellent tool for defining priorities and making sure that your actions are commensurate with risk. The prioritization matrix is a consensus method designed to rationally prioritize efforts on those projects that are most important to the organization. The strategic planning group can use this method to consistently prioritize improvement activities.

First, define your prioritization criteria such as patient safety, compliance, and efficiency. Some companies use additional prioritization criteria such as customer satisfaction (market share), cost, ease to implement, time to implement (e.g., quick hits). Then assign values for the impact of each potential project on the prioritization criteria (see Table 13.1).

Second, list the prioritization factors in the columns. The prioritization factors may be given a weight for their relative importance. Then list potential projects in the rows of the prioritization matrix. Assign ratings for each potential project for each prioritization factor. Potential projects are prioritized by multiplying each option's score by the weight for that factor and then summing up the products of all factors for a total rating (see Table 13.1).

Table 13.1 Prioritization Criteria Example

	Weight	High (9)	Medium (5)	Low (1)
Patient safety	10	Certain patient risk	Potential patient risk	Low patient risk
Compliance	7	Warning letter, Systemic major observations	Isolated 483 or major observations	Preventive compliance
Efficiency	3	Customer responsiveness impact	Moderate efficiency impact	Process efficiency impact
Market share	3	Increases long-term competitiveness and market share	Reduces hidden costs of lost sales	Little impact

The total rating is then used to rank projects. At this point, you can go down the list and assign as many projects as you have resources for. As one project is completed, the next one on the list is resourced.

WARNING

Once you have determined your priorities, do not change them without good reason. A call from the VP with a new pet project is not a good reason. It is a demonstration of individual gut feel and lack of transparency replacing an open and transparent, data-driven, consensus method. It is better to understand why this new project is viewed as important enough to displace other priorities. How does it rank against other projects in terms of feasibility, cost, and risk? It is better to understand the reason and adjust your model or analysis if necessary.

WARNING, WARNING

Do not continue to add projects that you do not have resources for. Consider it a red flag if management says to "Figure out a way to do it anyway." Lack of prioritization and misalignment with resources will result in dilution of your resources for the high-risk issues. Understand the need and adjust your model accordingly. Seek alignment with management. Allocate resources to high importance projects. It is for management with executive responsibility to provide adequate resources. If high-risk projects do not have enough resources, then escalate the risk to management review.

When priorities are defined, it is time to translate the quality objectives into action plans. Some approaches go straight to action plans and others break it down to more manageable steps. Action plans should provide all the relevant details of what needs to be done, expected results, the scope of the project, measurement criteria, team members and sponsors, other resources, milestones, etc. An A3 approach or Six Sigma charter are good examples. Implement action plans

Table 13.2 Prioritization Matrix Example

Quality and Compliance Prioritization Matrix

Prioritization Factors	Patient Safety	Compliance	Efficiency	Market Share	Ease to Implement	Total Rating
Weight	**10**	**7**	**3**	**3**	**3**	
Warning letter remediation	9	9	1	9	1	186
Process ownership and control	5	5	5	5	5	145
Design for quality	9	5	5	5	1	125
Manufacturing excellence	9	5	5	1	5	141
Focus on prevention	9	9	9	5	5	201
Shorten the feedback loop	5	5	5	5	5	185
Management responsibility training	5	5	5	9	9	165
Establish Big Q council	5	5	5	9	9	237
Integrated EQMS implementation	5	5	5	5	1	129
Big Data analytics	5	5	5	9	1	121
Other						
Other						
Other						
Other						

using good project management and problem-solving techniques. Good governance includes regular project updates and monitoring of progress.

Many projects stall because of cultural resistance to change. Consider change management (see Chapter 15, Alphabet Soup) principles to ensure stakeholder alignment for critical projects. Do not underestimate the importance of and effort required for change management.

Capability is the intersection of having capacity (enough resources) and competence (ability). Top medical device companies must know the skills and competencies of their organizations. Develop an organizational structure for the quality and compliance function, including needed resources to ensure required capability. Outside of the quality organization, promote the use of quality improvement teams, process excellence models, or centers of excellence for processes. Determine linkages or communication channels.

Get the right people on the bus for where the strategy is taking you. Build a strong team with the required capabilities. Define key roles and responsibilities for your quality and compliance organization. Identify key competencies for each role. Some typical, but not all inclusive, roles and responsibilities are included in Table 13.3.

Depending on the size of your organization, these may be specialized roles or require one person to wear many hats. You may need to supplement internal skill sets with occasional outside subject matter experts or consultants.

For each function, create a career ladder for growth and development. For example, a quality engineer might progress from associate engineer, to engineer, senior engineer, and principal engineer. A career ladder recognizes, and rewards increased subject matter expertise that comes with experience. Create a framework of competencies for each role. Do regular talent reviews to optimize and develop the overall organization.

In summary, key success factors for translating a vision into strategies and plans include

- A well-articulated vision and quality objectives
- Defined strategic goals
- Identifying and removing barriers to success
- Effective prioritization
- Employee resource allocation
- Key capabilities in place
- Clear roles and responsibilities
- Process performance measurement and control

A governance structure is necessary to ensure progress. Determine who and how you will govern the QMS. Obviously, this is dependent on the size structure and complexity of your organization. This could be your site management team or a formally designated quality council. It is necessary that the quality governance council has sufficient influence and formal authority to understand risks,

Table 13.3 Quality Organization Competencies

Analyst	Sources of data, management review, etc.
Auditor	Internal, supplier
CAPA specialists	
Change control specialists	
Change management resources	
Complaint handling unit, call center	
Document control specialists	
Improvement techniques and resources	Green belts, black belts in Six Sigma, root cause analysis
Inspection management	
Inspectors	
Issue (recall) management	Failure investigations
Project management resources	Project leaders, CAPA leaders
Quality engineers	Design, manufacturing, supplier
Quality system specialists	
Records management	
Regulatory	
Reliability/safety engineer	
Root cause analysis or problem solving	
Software quality engineer	
Statisticians	
Technicians	
Training specialists	

allocate resources, and monitor progress. The quality governance council can help to manage stakeholder alignment.

For each goal, consider the following:

In summary, develop a strategic plan to help you get from where you are to where you want to be. Some examples of steps and tools to use are given in Table 13.4.

In conclusion, it is essential to translate your vision and quality objectives into concrete strategies and plans. Quality planning includes the everyday, ongoing things you do to ensure quality, safety, and effectiveness of every product you make. And it requires a longer term view for continued improvement and success. Starting from your current situation, create your road map to the future. Identify the barriers to success. Determine your strategies to improve efficiency, effectiveness, and enable current activities. Define and plan for the necessary resources. This includes competencies and capacity to realize necessary capabilities. Prioritize projects to ensure actions are commensurate with risks. Manage projects and monitor results. It is that simple. It is the details that matter to ensure a suitable and effective QMS.

Table 13.4 Strategic Planning Steps and Tools

What	Tools
Plan to plan	Annual planning cycle
Environmental scan	Internal situation analysis
	External environment analysis
Visualize the future	Merlin exercise
	Define barriers to success
	Benchmarking
Understand strengths and weaknesses	SWOT analysis
	MEDICS maturity analysis
	Compliance profile, heat map
	Stakeholder analysis
	Barriers to success
	Quality situation analysis, status, quality metrics, management review, complaint analysis and trends, recall analysis, MDRs trends and analysis, market share, customer sentiment, etc.
Establish strategic quality objectives	Strategy
	Quality and compliance objectives
	Multiyear plans
Identify projects	Prioritization matrix
	Action planning road map
Identify resources	Gap assessment (people, skill sets, capabilities)
	QIPs (quality improvement project teams)
	Training gap assessment and analysis
	IT infrastructure
	Facilities, labs, supplies, computers, software, travel, etc.
Develop action plans	A3, 4 Up, project charters
	Training plans
	Change management plans
Implement action plans	Project management
	Six Sigma
	Change management
Track and measure progress	A3, 4 Up, project updates
	Scorecards
	Status updates and schedule
Evaluate results	Metrics
	CAPAs and effectiveness checks
	Results per project charter
Feedback loop	Annual quality survey
	Status updates
	Metrics
	Stakeholder alignment analysis

Alignment

Stakeholder alignment depends on knowing and working with your key stakeholders. Building an efficient and effective quality management system (QMS) is dependent on every person in the company. Quality and compliance leaders or management representatives cannot do this on their own. Often, the management representative has much responsibility but little direct authority over those responsible for quality and compliance outputs. Influencing stakeholders is therefore an important consideration. And influencing stakeholders requires alignment in goals and priorities.

Implementing, changing, and improving your QMS requires on-going effort to seek alignment with stakeholders in your organization. This becomes more and more important the higher you go in the organization. At lower levels in the organization it is important to be aligned on actions to execute tasks per the procedures of a defined QMS. At higher levels, it becomes more important to establish the totality of a suitable QMS.

In Chapter 3, Establish and Maintain, we defined an effective QMS as:

- The processes and documented procedures
- The organizational structure needed to implement those processes and documented procedures
- The information technology (IT) infrastructure necessary to perform processes, maintain and control records, and ensure compliance
- The personnel with appropriate skills, experience, and training to properly and consistently execute processes and procedure.

All of these elements need to work together in a coordinated manner. Misalignments between any of these elements can render the best strategy useless. It takes considerable effort to keep all of those elements aligned and moving in the same direction. In this chapter, we will discuss key areas of needed alignment and some common pitfalls.

The management representative and management with executive responsibility have a very important and unique relationship. Trust is an essential ingredient. Management with executive responsibility usually does not have sufficient depth of knowledge and experience to create a suitable QMS. Yet, they have responsibility to do so. Appointing an experienced, capable management representative is one of the most important decisions they must make. Smart executive management realizes that their management representative is the one that can "keep them out of regulatory jail." Building a relationship of trust requires on-going dialogue about the business situation and the challenges and opportunities that it brings. It requires clear expectations and alignment on tolerance for risk. There are never enough resources for everything which results in the need for risk-based choices. The changes to ISO 13485:2016 make it abundantly clear that actions must be commensurate with risk. The management representative and management with executive responsibility need to stay closely aligned regarding risk tolerance and providing adequate resources for a suitable and effective QMS.

An efficient and effective QMS requires an organizational structure that facilitates efficiency and effectiveness. Make sure there is alignment between company organizational structure and the main processes that are defined in your quality manual. Make sure all functional organizations understand which regulatory processes they are specifically responsible for and those that influence them. Use informal or formal (e.g., RACI model for **R**esponsible, **A**ccountable, **C**onsulted, **I**nformed) methods to ensure on-going alignment on processes. See Chapter 15, Alphabet Soup, for details on the RACI model.

Build working relationships with the functional leaders or process owners. A process approach should be part of your management strategy. A process approach is used to efficiently and effectively manage and control the processes that make up their organization, the interaction between those processes, and the inputs/outputs that tie those processes together. To be well managed, a process must have:

- An owner
- Regulatory requirements are all identified and included
- Be defined in procedure(s) and work instructions with adequate detail
- Have appropriate management infrastructure (i.e., IT systems) in place (if applicable)
- Measurements and control points are defined
- Information to manage, control, and monitor the process in real time
- Demonstrate stable, predictable, repeatable performance and yield consistent results and records
- Have effective mechanisms for communication with suppliers and customers, both internal and external (reference SIPOC model)
- Improve performance as necessary.

One of the most frequent causes of an inefficient and ineffective QMS is lack of alignment with functional or process owners. Most process owners are not well educated about QMSs (but you can give them a copy of this book). They simply do not understand the importance and relevance of their roles from a quality and compliance standpoint. Without constant management, they typically drift back to their core knowledge, neglecting process management and control. Process owners sometimes incorrectly perceive the quality organization as being responsible for quality and compliance. Help them understand the relevance and importance of their role (ISO 13485:2016 clause 6.2 (d)). It requires on-going stakeholder management to stay aligned on roles and responsibilities. The higher in the organization you are, the more important this becomes.

Set guiding principles for process ownership. Provide training for process owners. You must move them from a state of unconscious incompetence to unconscious competence. Help process owners to understand the importance and relevance of their role in establishing an effective and efficient QMS. Leverage learning moments.

Look for alignment of different functional groups with each other and the resulting impact on quality and compliance results. For example, the intersection of functions such as R&D, operations, marketing, and quality/compliance/regulatory is critical to new product development. For most companies, new product development is one of their most significant investments of resources. These companies go to great lengths to create a business process (e.g., product development process) to provide governance, set priorities, determine structure, and allocate resources for product development. They create stage gates and set expectations for Return on Investment (ROI), Cost of Goods Sold (COGS), and so on. Effective organizations also build the regulatory requirements (e.g., design control) into that business process.

Look for signs of misalignment such as process owners wanting to keep the regulatory requirements (e.g., design control) separate from the associated business process (e.g., product development process). Usually, this stems from a business leader/process owner that does not fully understand, value, and respect the regulatory requirements. They want to keep control of the things they value (such as designing the product) separate from the things they do not value (such as the perceived bureaucracy of design control). Sometimes they want to keep things separate because the regulatory requirements are outside of their core expertise or comfort zone. This creates dual processes that erode both effectiveness and efficiency. For example, they have one project plan that includes all business activities, financial plans, and business stage gates. They have another design and development plan for design control requirements only. They keep things separate because they do not want the FDA to see business information. Using two plans is more inefficient and duplication of effort. Even more importantly, it reduces effectiveness making it more difficult to plan well, see interrelationships and prerequisite activities, share information, and highlight priorities. It creates additional opportunities for error. And it causes intense frustration with employees who feel the consequences of duplication of effort.

The reality is that the FDA is not really interested in (and has no authority over) the business information such as pricing, ROI, or COGS. They only want to see that you are complying with the regulations and that results in safe and effective products. It is more efficient and effective to avoid dual processes. And if it is ever necessary to show business information to the FDA, it is acceptable to redact pricing or business information. Try to get to the heart of the matter and understand why process owners want to maintain dual systems with the resulting inefficiency and ineffectiveness.

Management representatives must be watchful for signs that misalignment is resulting in unacceptable product quality. It is important to understand the risk and escalate if necessary.

CASE STUDY

Lack of alignment can create insurmountable barriers to success. In several real-life cases, I have seen organizations attempt to design new products to address serious quality problems. Because the organization was designing a new product, the marketing organization wanted to add a few bells and whistles. Then operations decided to add some design for manufacturability requirements or cost of goods improvements. The product development team agreed to those design requirements. After some time and a lot of work, they came to the realization that it was not possible to meet all the design requirements. It was impossible to technically address the quality problem, add the features desired by marketing, and reduce cost or improve manufacturing capability. There were disagreements and long, drawn-out design review meetings. There were difficult stage gate meetings where management told the team to go back to the drawing board. The product development team was disillusioned but tried again. The cycle repeated several times with no measurable progress. Another project leader was put in charge. The old CAPA was closed and a new one opened. But the team was never able to create a product that could meet all of the design requirements. Management was frustrated, and the team was frustrated. None of the organizations were willing to compromise and change their requirements. The functional organizations could not get aligned on a critical project. The cycle repeated itself several times and the problems were never resolved. A serious product quality issue was left unresolved after years of frustrating effort. Does any of this sound familiar to you?

In the meantime, other parts of the organization were dealing with high levels of complaints and medical device reports. Several recalls were necessary. The FDA did an inspection and clearly cited the consequences of a serious quality issue. They documented the ineffectiveness of the CAPA system. The inability to deal with a long-standing quality problem raised questions about the suitability of the QMS. And because management is responsible for ensuring a suitable QMS, it raised concerns about management responsibility. The company got a warning letter.

Make sure there is alignment on goals and priorities between functional groups. When lack of alignment is evidenced, understand the risks and escalate to higher levels of management. Use patient safety to prioritize actions and break the gridlock. Actions must be commensurate with risk. Any issues concerning product quality in the field must be escalated if necessary. Management with executive responsibility must pay very close attention to CAPAs related to product quality, corrections and removals, class I recalls, or medical device reports!

Look for good alignment between processes. Remember that every process has inputs from suppliers (not just external suppliers, but internal processes as well) and outputs to customers (not just external customers but also internal processes as well). Every process owner must make sure they are aligned with the owners of connecting processes. Failure to do so will result in gaps (ineffectiveness) or overlaps (inefficiency) in requirements. Gaps in requirements are nonconformities. Overlaps can also create more disagreement and potentially, more discrepancies and opportunities for error. This is a very common source of quality and compliance problems as well as frustration of employees.

Make sure there is alignment between processes and supporting IT infrastructure. The IT system must support the process flow. It should enable mistake proofing and allow real-time control and monitoring. It must handle the multitude of records the quality system requires. It must allow collaborative environments and sharing of information. How will it be fully integrated so that you can "connect the dots"? For each step in your improvement strategy, determine how IT systems will support and enable it. Clearly define the impact of IT systems on process efficiency and effectiveness. We are in the fourth industrial revolution right now. Watch this space!

Check for alignment with all external suppliers. Make sure there are systems for sharing and controlling purchasing data (21 CFR 820.50 (b)). Make sure there are clear requirements for suppliers to provide notification of changes and seek alignment. Make sure there are mechanisms for performance monitoring, review, feedback, and CAPA if necessary.

Check for alignment with all customers. ISO13485:2016 clause 8.2.1 requires "As one of the measurements of the effectiveness of the quality management system, the organization shall gather and monitor information relating to whether the organization has met customer requirements." But, do not do it just because it is a regulatory requirement. Do it because it is vital to overall business success. Do it because it is the right thing to do.

Our DNA is as a consumer company—for that customer who's voting thumbs up or thumbs down. That's what we think about. And we think that our job is to take responsibility for the complete user experience. And if it's not up to par, it's our fault, plain and simply.

Steve Jobs

Build the quality and compliance team. Consider your own strengths as you determine the skillsets that you need in your team. A wise manager told me once that she always wanted to hire people that were smarter than she was. Make sure the skills and capabilities in your organization are aligned with your strategy. If your strategy is to develop products with a new technology, make sure the skill sets of your organization can support that strategy.

Your strategy should include a communication plan. Good communication and deliberate actions are necessary for on-going alignment. Alignment depends on a system of flowing down goals and objectives from one layer of management to the next. From the CEO to all individual contributors, everyone must know the quality policy, plans, and how their individual goals and objectives support that. Work with your human resources (HR) department to create a system to cascade goals and objectives throughout the organization.

Your quality policy and objectives should have a ripple effect much like tossing a rock into a pond. The ripples should spread and touch each person in your organization. If your quality policy and objectives do not have this effect, then your communication plan and objectives cascade are not working.

Make sure that reward and recognition systems have reasonable expectations for quality and compliance, based on the level and individual role. Process leaders should be evaluated based on process performance. Individual contributors should be evaluated on individual performance.

WARNING

I must also provide a caution about alignment given the context of a QMS for medical devices in a highly regulated industry. Alignment is about creating agreements, alliances, and a common direction. Alignment is a necessary ingredient for optimizing your QMS. But, do not let your quest for alignment become your weakness. Do not let it become your Achilles heel. Do not let your desire for alignment make you sacrifice your own standards of quality and compliance. Do not let your desire to be liked cause you to compromise your need to do the right thing. This is a very real danger. Quality and compliance personnel have to walk a very fine line.

Stakeholders should be identified as you create your monitoring system and balanced scorecards. It is important to ensure you accurately define the key stakeholders and have metrics that drive satisfaction. Monitor these metrics so that you understand how well you are aligned with their expectations.

Make sure metrics are aligned with goals! Remember Tom Peters' guidance from *In Search of Excellence*, "What gets measured, gets done." If measurements are not aligned with goals, you will get the wrong results. Review the monitoring and measurement capabilities in Chapter 8, Capabilities and MEDICS for an Effective QMS.

All these types of misalignment will reduce the effectiveness and the efficiency of your QMS. Misalignment creates uncertainty about roles and responsibilities. Because it is not clear who is responsible for an activity, it may be neglected. Or, if two organizations think it is their role, there will be duplication of effort. There may be disagreement about the ways things should be done. There will be inconsistency in the way it is performed. Mistakes will be made. There may be remaining disagreement that needs to be resolved. Serious misalignment may even lead to unacceptable behaviors and risk-taking. Stakeholders must watch for signs of misalignment:

- There is confusion, gossip, and frustration about processes and procedures.
- Silos form around individuals and departments. Turf wars occur.
- Comments like "I don't know, I just work here" are heard.
- There is an attitude of "That's not my job."
- New initiatives are seen as the flavor of the month.
- More negative behaviors are seen as transparency, openness, and trust disappear.

WARNING

Watch out for misalignment between authority and responsibility. Both ISO 13485:2016 and the quality system regulation have clear expectations for authority and responsibility. Yet, many companies make the mistake of assigning responsibility to individuals without giving them appropriate authority or the resources to get things done.

CASE STUDY

A VP of Quality and Compliance gave the audit organization total responsibility for compliance at the company. The director of the audit organization only had the true authority and resources to conduct audits and report the results. The audit organization did NOT have the authority or the resources for corrective and preventive action for audit observations. Even more importantly, to have the auditors responsible for the corrective and preventive action would be a conflict of interest and diminish their independence.[1] Despite much discussion and frustration, the disconnect was not resolved. The audit organization felt powerless and frustrated. The process owners felt free to do whatever they wanted and did not address audit observations.

[1] Independence is a regulatory requirement and essential for effective audit

---

Alignment requires on-going effort. Check in with your stakeholders frequently. Use the MEDICS to keep your finger on the pulse of the organization. Look out for signs of misalignment. Use change management and stakeholder alignment techniques if signs of misalignment appear. Repeat as necessary.

The higher you go in the organization, the more important stakeholder alignment becomes. As a result, you will spend more of your time on stakeholder alignment. Influencing skills become more and more important. Every quality/compliance leader should give some thought to how they are spending their time. Then do a self-check to see that it is aligned with your quality strategy and personal goals.

In conclusion, it takes an on-going and concerted effort to ensure alignment of all parts of the QMS. Processes, process owners, and process enablers all need to be aligned to work in a common direction to avoid gaps and overlaps. Stakeholder analysis and alignment are important considerations in improving QMS efficiency and effectiveness.

PART V SUMMARY

Part V covered the basics of how to translate regulatory requirements and quality objectives into an integrated quality system. A process approach is a basic element of an efficient and effective QMS. Understanding the current business situation and strengths, weaknesses, opportunities, and threats is the starting point for developing a vision, strategy, and plans for improvement. Plans need to be detailed. Decisions about priorities need to be made. On-going stakeholder alignment is required to move the strategy forward resulting in an efficient and effective QMS.

Long-term success to improved QMS effectiveness and efficiency is evidenced by:

- Reduced external regulatory observations
- Reduced cost of poor quality
- Increased value of quality and impact to the business
- Proactive investments in quality are well accepted
- Quality is valued as a competitive strength and market differentiator
- Compliance leads to improved quality and customer satisfaction.

Improvement tools and techniques

Alphabet soup

CAPA

In Chapter 8, Capabilities and MEDICS for an Effective QMS, the concept of CAPA and improvement were introduced as one of the key capabilities or MEDICS (Monitor, Embrace, Define, Identify, CAPA, Share) that every medical device company must have. This chapter delves more deeply into the concept of CAPA and useful tools and techniques. Corrective and preventive action (CAPA) is the heart of an efficient and effective quality management system (QMS). Without a rigorous CAPA process, a medical device manufacturer does not have the capability to prevent problems or improve. CAPA should be considered the single most important process within your QMS. The CAPA process is a key factor in the health of your QMS. Without a robust CAPA process, it is impossible to have an efficient and effective QMS.

CAPA can be much more than just a fulfillment of regulatory requirement. If well implemented, it can be the tool that drives measurement and analysis, keeps management informed, and mitigates risk.

TIP

Keep your overall strategy in mind as you develop your CAPA process. Your CAPA process should, over time, help you to become less reactive and more proactive. As your QMS matures, you should be focusing more resources on preventive action than corrective action. And a good CAPA process must be aligned with a focus on improved quality outcomes.

CAPA is a rigorous, closed-loop system for identifying, analyzing, and improving products and process problems. Closed loop means it has a feedback loop to verify that actions were effective. The FDA (Food and Drug Administration) regulations require that medical device manufacturers establish and maintain systems to:

- Analyze processes
- Investigate cause of nonconformances
- Identify actions
- Verify or validate actions are effective
- Implement and record changes
- Disseminate information
- Submit relevant information for management review.

Medical Device Quality Management Systems. DOI: https://doi.org/10.1016/B978-0-12-814221-9.00015-4

Some key definitions for understanding:

- Correction—action to eliminate a detected nonconformity (e.g., disposition product as reject (scrap), rework, repair, release)
- Corrective action—action to eliminate the cause of a detected nonconformity or other undesirable situation. Corrective action is taken to prevent recurrence.
- Preventive action—action to eliminate the cause of a potential nonconformity or otherwise undesirable situation. Preventive action is taken to prevent occurrence.

> **NOTE**
>
> The FDA and ISO 13485 both use the word "preventive" in the regulations. Although "preventative" is commonly used in industry, it is inconsistent with the wording in the regulations. The regulations use the term "preventive," not "preventative"!

ISO 13485:2016 addresses CAPA under clause 8.5 Improvement. It breaks CAPA into two subsections, one for corrective action and one for preventive action, which is a useful distinction from the FDA wording. This makes it clear that not every corrective action has a corresponding preventive action. This is consistent with the expectations from the GHTF (Global Harmonization Task Force) guidance on corrective and preventive action that states that "the acronym CAPA will not be used in this document because the concept of corrective action and preventive action has been incorrectly interpreted to assume that a preventive action is required for every corrective action. This document will discuss the escalation process from reactive sources which will be corrective in nature and other proactive sources which will be preventive in nature." Regardless of the nature of the source, the CAPA process will be similar and we will use the acronym CAPA in this book. Additionally, preventive action from proactive sources is discussed further under risk management in Chapter 8, Capabilities and MEDICS for an effective QMS.

Many companies struggle to make their CAPA systems efficient. They put every problem into the CAPA systems and everything stalls. This is a very common mistake. Companies do not recognize that they can take a risk-based approach to handling CAPA. The Preamble to the Quality System Regulation, comment 159, indicates that "the degree of corrective and preventive action taken to eliminate or minimize actual or potential nonconformities must be appropriate to the magnitude of the problem and commensurate with the risks encountered." They go on to say that "the FDA does expect the manufacturer to develop procedures for assessing the risk, the actions that need to be taken for different levels of risk, and how to correct or prevent the problem from recurring, depending on that risk assessment." So, all medical device manufacturers should create their own risk criteria for their CAPA process. And they should use those criteria consistently.

The concepts of CAPA are linked to the expectation to determine the significance and risk of nonconformities. Nonconformities are defined as nonfulfillment of a requirement. These nonconformities may be related to product, process, or the QMS. Based on the associated risk, the manufacturer may make a correction and also a corrective action to prevent recurrence. In cases where a nonconformity has not yet occurred, but there is evidence of a change or trend, then a preventive action may be considered. For example, when monitoring acceptance data, if there is a trend indicating the control limits may soon be exceeded, then a preventive action may be necessary.

Not every nonconformity needs to go into the CAPA system. Make sure you investigate to understand the scope, extent, criticality, and systemic nature of a nonconformity. Use your risk criteria to determine if a nonconformity requires no action, further monitoring, correction, corrective action, preventive action, or additional remediation.

Consider the steps in Table 15.1 in creating your CAPA program.

A good CAPA process requires defined sources of information. You must identify the "feeder processes" for CAPA. The FDA regulation explicitly identifies the following sources of data:

- Processes
- Work operations
- Concessions
- Quality audit reports
- Quality records
- Service records
- Complaints
- Returned product
- Other sources of quality data

Additional examples of data sources are as follows:

- Nonconforming material or product
- Equipment maintenance, calibration, adjustment
- Process controls
- Environmental controls
- Acceptance testing
- Medical device reports (MDRs)
- Recalls
- Design controls
- Production and process controls
- Purchasing controls
- Management review.

Of these sources, complaint handling is one of the most important. In fact, in the FDA's *Guide to Inspections of Quality Systems* (usually referred to as QSIT

Table 15.1 Steps for Creating a CAPA Program

	What	Why	How
Planning	Identify internal and external sources of data that are indicators of process and product performance.	Data sources aligned to business goals and quality objectives will receive appropriate attention.	Annual planning, goals and objectives, quality plans, CAPA SOP (standard operating procedure).
	Establish sources of data.	Internal and external sources of data are needed to monitor product quality and the health of the QMS	List sources, processes, measurement criteria.
			Identify resources and responsibilities for each data source and data elements.
	Determine data elements.	Data elements provide information regarding nonconformities and the effectiveness of the established processes within the data sources.	For each data source, define data elements (quantity, rate, status, etc.).
	Develop risk criteria for escalation.	Actions should be commensurate with risk. Ensure adequate resources for higher risk issues.	Risk categories, escalation criteria (such as action or trigger levels). Document escalations.
	Ensure software for CAPA system, measurement, and analysis is validated.		Software validation protocol and report.
Measurement and analysis	Perform measurement, monitoring, and analysis processes.	Measurement is a set of operations to determine a value of a data element.	Measure per defined method, frequency, precision, and accuracy.
			Measurement data should be retained as a quality record.
		Monitoring is systematic and regular collection of a measurement. Determines conformity or nonconformity of processes and products	Conduct monitoring per defined procedure. Maintain results as a quality record.
	Use appropriate statistical techniques.	Ensure that measurement, monitoring, and analysis are valid and meaningful.	Per established procedure for valid statistical techniques. Use trained resources.

(Continued)

Table 15.1 Steps for Creating a CAPA Program *Continued*

	What	Why	How
Improvement	Investigate.	Analyze within and between sources of data. Determine scope and extent of issue.	Example: Is a failure at finished goods acceptance testing related to complaints? Create an investigation report.
	Root cause analysis.	Eliminate the cause of a nonconformity.	Root cause analysis tools.
	Identify actions.	Determine need for no further action, additional monitoring, correction, corrective action, preventive action, remediation, and/or escalation.	Document planned actions, resources, and timelines.
		Determine other actions (e.g., identification of resources, regulatory submissions, timelines).	Project management tools
		Methodical, rigorous process improves results.	Six Sigma or process excellence tools provide rigorous methods.
		Determine required outcome. Predetermined criteria for success ensures expected results are achieved.	Identify criteria for effectiveness checks.
	Verify and validate.	Verifying or validating actions are effective prior to implementation.	Protocol and report.
	Implement and record changes.	Put plans and actions into place.	SOP or specification updates, training, and change control.
			Ensure appropriate notification/ communication of changes per your change control process
	Check effectiveness.	Ensure actions are effective and do not adversely affect the finished device.	Gather data over a period of time to ensure actions were effective against predetermined criteria for success.

(*Continued*)

Table 15.1 Steps for Creating a CAPA Program *Continued*

	What	Why	How
Communicate	Disseminate information.	Process owners must be aware of measurement, monitoring, analysis results and CAPA progress. Quality organization (or other designated individuals) should be aware of issues and CAPA progress.	Share information with relevant process owners, subject matter experts, and those directly responsible for ensuring quality. Meeting minutes, approved quality records.
	Management review.	Relevant information should be submitted to ensure actions are commensurate with risk	Safety-related issues should be escalated immediately.
		Management responsibility to ensure a suitable QMS. Data should be appropriate to the scope of management responsibility	Management review meeting minutes, decision, action items.
		Good data leads to informed decisions, continuous improvement, and customer satisfaction.	

or Quality System Inspection Technique), complaint handling and medical device reporting are bundled in a "satellite" with CAPA as one of the four main subsystems covered in inspections.

Medical device manufacturers typically face the following challenges with respect to CAPA:

1. It is challenging to track CAPA activities through the process. There is lack of visibility to open items such as approvals and actions completed on time.
2. It is difficult to monitor the status and progress of individual CAPAs. There is difficulty gathering metrics on the CAPA process.
3. It is easy to forget to measure effectiveness at some point in the future.

An automated management system for CAPA can be very helpful to deal with these challenges. There are several good versions out there. Note: Be sure you have an effective and efficient process before you automate it! The automation will help you ensure documentation of all notifications, approvals, and actions. It can also help you to house or cross-reference documents that might be housed in other areas such as complaint or change control files. And it can provide data for monitoring your overall CAPA process and individual CAPAs. Do not forget to validate the software! If you're a growing company, consider switching to an automated system sooner rather than later because data migration becomes more difficult the longer you wait.

TIP

Always notify the IT organization of FDA inspections. You do not want your CAPA IT system (or complaints, calibrations, etc.) to be down for planned maintenance during an FDA inspection. Additionally, they may be needed to answer questions regarding validation of the system.

Be mindful of the following warning signs that indicate problems with your CAPA process:

- Recurring issues
- Inability to manage sources of data and understand early trends and issues
- More reaction than prevention
- Resources (people, money, time) are spent more on dealing with failure than preventing it.
- Everything goes into your CAPA system and stalls.
- CAPA is perceived as a bureaucratic process owned by the quality organization.
- Management does not want to open a CAPA because they do not have enough resources.
- There is a perception in your organization that "There's never enough time to do it right, but there's always enough time to do it over."

WARNING!

One of the most egregious comments that I have heard from senior management was "Do not open a CAPA if you do not have enough resources to deal with it." This violates the basic concept of management responsibility for ensuring adequate resources for a suitable QMS!

Tips for a successful CAPA process:

1. Create a cross-functional team, sometimes referred to as CAPA board or management review board (MRB), to review and discuss issues that may be candidates for a CAPA. This board should meet regularly to review inputs and CAPA requests, assess issues, and make determinations to open a CAPA. The board should review a listing of CAPAs, including resources, status, and acceptable progress. Ensure proper documentation of meetings including an agenda and meeting minutes.

2. Risk management and prioritization should be embedded in your decision-making process for CAPA. Consider risk to determine if a nonconformity requires formal investigation, initiation of a CAPA, and prioritization and urgency of CAPA activities. Risk-based decisions should help you reduce the total volume of CAPAs while ensuring that CAPAs that *are* initiated are treated with appropriate rigor, urgency, and resource allocation.

3. Good root cause analysis (RCA) is *essential* to effective CAPA. Without a good RCA, you are just wasting precious time, money, and resources doing things that are not helpful. Good RCA takes time, effort, and skilled resources. We will discuss some RCA tools in this chapter, but this is a very large and deep topic. It is essential that you develop a toolbox of RCA tools and have trained, skilled resources on your CAPA teams!

4. Data elements should help you understand the process performance of each data source. For example, for complaint handling consider data elements such as quantity and/or rate by product or product family, customer, reason, complaint codes, severity, and status of each compliant file. Refer to Chapter 9, Compliance Must Result in Improved Quality, on metrics for additional information.

5. Make sure you have metrics on your CAPA process itself to monitor the performance of the CAPA process. These metrics should be communicated in management review. Make sure your metrics address effectiveness, efficiency, and cycle time. Consider metrics such as number of CAPAs, number or percentage of CAPAs that have met key milestones (e.g., initiation, investigation complete, actions complete, and effectiveness check complete), percent of CAPA that were closed as effective, aging of CAPAs, etc. Note: There are always some very old, moldy CAPAs that needed to be highlighted and addressed. If you only measure average CAPA age, they may get lost. Consider using average and median age, or average and 90th percentile age, or number of CAPAs older than 1, 2, or 3 years.

6. Make sure you have a system for communication and sharing of information. A CAPA data management system can be very helpful by providing communications, requiring approvals, and prompting activities according to required timelines. It can send reminders including escalation of issues if necessary. Without a CAPA data management system, you must rely on good project management skills.

7. Project management skills can be very useful for dealing with significant and complex problems. Consider it a good investment to have a project manager for your most high-risk CAPAs. Complex problems require good project management and adequate resources to address them effectively. Many companies use the concept of a "CAPA specialist" to coordinate and manage significant CAPAs. Your most important issues deserve the structure of project management.

8. Create expectations for what documents are maintained within the CAPA system or other systems such as complaint files and change control records.

Create expectations for where things will be filed and how references will be created. This will prevent confusion and missing records in the future.

9. Ensure that your corrective and preventive actions are not limited to only the known or identified nonconformity. If one nonconformity was found within the sample reviewed, are there additional nonconformities not identified that should be? For example, your internal audit program looked at 58 complaint records and found one case where a complaint was incorrectly classified as nonreportable (MDR). Do not limit actions to just correcting that one known case. Make sure you investigate further to understand the true error rate and correct ALL complaint files with the same error. **Failure to look broadly is one of the key reasons that problems known internally are also discovered later by regulatory authorities.**

DEALING WITH INDIVIDUAL CAPAS

The steps for dealing with individual CAPAs are as follows:

1. Identify and prioritize.
   a. Define the problem.
   b. Notify appropriate personnel.
   c. Create a CAPA request.
   d. Review to determine if issue meets risk criteria and warrants a CAPA.
   e. Conduct failure investigation, if appropriate, to determine scope, risk, and escalation.
2. Investigate.
   a. Identify any immediate actions (e.g., recall or product-hold order to control distribution of nonconforming product).
   b. Conduct RCA.
   c. Plan effectiveness check.
3. Review and approve plan.
   a. Identify actions and assign resources, including cross-functional team.
4. Verify and/or validate actions.
5. Implement actions.
   a. Use the change control process.
6. Conduct an effectiveness check.
   a. Measure against predetermined criteria
   b. Monitor to ensure no recurrence of the problem.

CAPA INSPECTION PREPAREDNESS

Because an effective CAPA process is so important, it is always inspected by the FDA. Your CAPA process will always be evaluated during ISO audits. Yet, year after year, we see FDA Form 483 inspectional observations for lack of or inadequate CAPA procedures.

The FDA's QSIT provides details on how the FDA inspects the CAPA subsystem. The QSIT guide indicates that "One of the most important quality system elements is the corrective and preventive action subsystem." It is imperative that a medical device manufacturer has an efficient, effective CAPA process and that it is ALWAYS inspection ready.

Common Form 483 and warning letter citations for CAPA include the following:

- The procedure does not adequately define the process for executing a CAPA.
  - It does not require use of appropriate statistical methodology.
  - It does not require verifying or validating corrective or preventive action.
  - It does not require analyzing quality data.
- RCA (Root Cause Analysis) was not conducted.
- Effectiveness checks were not conducted.
- The procedure does not include customer complaints as a source of data.
- Documentation problems such as:
  - There were no analysis data.
  - Neither the investigation nor the implementation of corrective actions was documented or referenced in the CAPA document.
  - The CAPA was closed with no corrective action or effectiveness check.
  - There was a failure to ensure that information related to quality problems is disseminated to those directly responsible.
  - The results of the failure analysis are not transmitted to QA in a timely manner.

The FDA's approach to inspection of the CAPA subsystem is clearly defined in the QSIT manual. Do not be surprised. CAPA process owners and specialists should read the QSIT manual in its entirety. CAPAs related to product, process, and quality system are all within the scope of FDA's investigation of your CAPA system. The QSIT manual defines the following inspectional objectives for CAPA:

1. Verify that the CAPA system procedures address the requirements of the QSR (Quality System Regulation) and have been defined and documented.
2. Determine if appropriate sources of product and quality problems have been identified. Confirm that data from these sources are analyzed to identify existing product and quality problems that may require corrective action.
3. Determine if sources of product and quality information that may show unfavorable trends have been identified. Confirm that the data from these sources are analyzed to identify potential product and quality problems that may require preventive action.
4. Challenge the quality data information system. Verify that the data received by the CAPA system are complete, accurate, and timely.

5. Verify that appropriate statistical methods are employed. Determine if results of analysis are compared across different data sources to identify and develop the extent of the product and quality problems.
6. Determine if failure investigation procedures were followed. Determine if the degree to which a quality problem is investigated is commensurate with the significance and risk of the nonconformity. Determine if failure investigations are conducted to determine root cause. Verify that there is control for preventing distribution of nonconforming product.
7. Determine if appropriate actions have been taken for significant product and quality problems identified from data sources.
8. Determine if corrective and preventive actions were effective and verified or validated prior to implementation. Confirm that corrective and preventive actions do not adversely affect the finished device.
9. Verify that corrective and preventive actions for product and quality problems were implemented and documented.
10. Determine if information regarding nonconforming product and quality problems and CAPA have been properly disseminated including dissemination for management review.

The data source most frequently inspected is complaints. And from there, the FDA will also inspect those that are reportable to the FDA as MDRs. You should be prepared to provide a list of your complaints. Give some thought ahead of time on how you will provide that list. It may be a printed list. But, sometimes the investigator will want to watch someone using your data system live. Sometimes the investigator will ask for an Excel spreadsheet on a CD or flash drive. With this data, they will do their own analysis of complaint data. Some of the FDA investigators are very skilled with pivot tables and analysis and will return the next day with very specific questions and follow-ups.

QSR Section 820.198(c) requires that for any complaints (or other indicators such as service records) that involve a possible failure of a device, labeling, or package, there should be a failure investigation. Nonconforming product 820.90 (a) also requires determination of need for an investigation. A good failure investigation includes:

- Unique identification (title and number)
- Description of the problem and identification of failure modes
- Severity and occurrence rate of the failure
- Impact (health hazard analysis) of the failure if applicable
- Details on failure distribution, location, customers, geographies, etc.
- Determination of the significance of failure modes (show linkage to tools such as risk analysis)
- Rationale for determining if a CAPA is required
- Controls for preventing distribution of nonconforming product
- Investigation method, procedures, tests, inspections
- Tools used in the investigation (gages, equipment)

- Description of examined documents, components, parts, labeling, processes, etc.
- Results and analysis methods
- Conclusions, reports
- Formal acceptance and closure.

WARNING!

Appropriate statistical techniques must always be used. If the statistical tools and techniques used are not valid, then all analysis and conclusions will be invalid as well. Garbage in, garbage out. Nonstatistical techniques will include reviews by quality boards, approvals, and other methods.

CAPA INSPECTION PREPAREDNESS CHECKLIST

Your CAPA process should always be inspection ready! But, if the FDA comes in tomorrow, here are some last-minute things you can do to put some "lipstick on the pig":

✓ If possible, determine the scope of an FDA inspection—is it a routine level 1 or level 2 inspection? For a routine inspection, the FDA will most likely look at 2 years of CAPAs (or possibly CAPAs since the last inspection you had).

✓ Be prepared to show a listing of CAPAs for the identified time frame. If you are printing a systems-generated list, plan ahead for what data fields you will include in the listing, for example, CAPA number, title, initiation date, type (product, process, QMS), status (open, waiting effectiveness check, closed), closure date. Investigators may ask for additional detail, but this is a good starting point.

✓ In a "for cause" inspection or because of a recall, the investigator will also look at the CAPAs for those specific issues. Tip: Your CAPA files for recalls must be rock solid.

✓ Be prepared to show routine analysis reports, with conclusions, for all sources of data. Be prepared to discuss any trends or outliers.

✓ Be prepared to show what and how statistical techniques are used for analysis of sources of data. It is important to show consistent methodology across sources of data.

✓ Tip: When the investigator requests a sample of files to review, present all files in a consistent, uniform manner. Remember, the expectation for "establish" includes **consistently** executing to your documented procedures. Show consistency and thoroughness.

✓ If you receive advance notice of an inspection, you may have time to identify high-risk CAPAs and review individual files for adequacy, completeness, clarity, and potential issues. Your review should include the following:

- Is there an investigation summary that provides a clear summary of the investigation?
- Are CAPAs for corrections and removals clear and complete? Is the timeline for investigation and risk assessment reasonable and clear?
- If applicable, is the health hazard evaluation complete and consistent with the rest of the CAPA?
- Were all changes (design, process, etc.) made adequately verified, or where appropriate, validated and documented?
- Is the sequence of events/flow of information understandable and logical?
- Are actions and timelines commensurate with the risk?
- Review all action items and commitments for evidence of closure on time.
- Are there any late or overdue actions? Are there any significant time gaps or lack of reasonable activity?
- Have root causes been clearly determined with RCA methods identified? Is there documentation to support the root cause(s) as well as cause(s) that have been eliminated?
- Do the corrective actions clearly address the root cause(s) and contributing cause(s)?
- Are the criteria for effectiveness checks clearly identified, including predetermined criteria for success? Are the criteria for effectiveness reasonable (not too easy, not too hard)?
- Does the CAPA documentation all stand on its own or will it require explanation by a subject matter expert (SME)?

✓ Be prepared to provide documentation on software validation for your IT systems, spreadsheets, or database.

As your CAPA process matures, you should be rewarded with a reduction in quantity and severity of issues. You will shift your resources more and more from reactive to preventive actions. Better processes and products should lead to improved customer satisfaction. And that is good for your business.

IMPROVEMENT TOOLBOX AND LEXICON

An effective and efficient quality management needs a comprehensive toolbox. The skill is choosing the right tool for the need. If your only tool is a hammer, then everything looks like a nail. The tools presented below are classic quality management tools that every quality and compliance leader in a medical device company should know about. These are tools that every management representative should be aware of. Any engaged manager with executive responsibility should have basic understanding of these tools in order to ask the right questions. And most FDA investigators are familiar with and understand them.

ROOT CAUSE ANALYSIS

In 1986, the space shuttle Challenger, on its 10th mission, broke apart 73 seconds after launch, killing all seven crew members. The breakup began after a joint in the right solid rocket booster failed at lift-off. As most people already know, this was caused by failure of the O-ring seals used in a joint that were not designed to handle the unusually cold conditions for this launch. Based on this quick RCA, a redesign of the O-rings seems like a very reasonable corrective action. But, that would be a very naïve and incomplete approach.

We must dig deeper and understand why the O-rings were a problem. In reality, the Challenger disaster resulted in a 32-month break in the shuttle program and the formation of the Rogers Commission appointed by President Reagan to investigate the accident. The commission determined that NASA's (National Aeronautics and Space Administration) organizational culture and decision-making process were contributing factors in the launch. NASA managers had known for some time that the O-rings from the contractor, Morton-Thiokol, contained a catastrophic flaw. Yet, they failed to heed warning from engineers not to launch in the unusually cold temperatures that morning with ice covering the launch pad. So, what is the real root cause:

- The design of the solid rocket boosters?
- The Challenger being launched outside of approved operating conditions?
- The solid rocket boosters were engineered by the low bidder?
- The failure to listen to warnings from engineers?
- The perception that, after 10 successful missions, launches are routine?
- The poor decision making or unacceptable pressure to launch?
- Group thinking allowing a known problem to be considered an acceptable risk?

In the example above, if the only root cause identified was the O-rings, then fixing the O-rings would be the only corrective action taken. There would be real danger of another major problem in a future shuttle. Good root cause analysis involves identifying the primary and contributing causes of a problem or undesired event. Problems are best solved by removing the true root cause(s) rather than the obvious symptoms. This leads to permanent resolution and prevention of other problems. RCA is aimed at identifying the layers of causes and contributing causes.

TIP

Root Cause Analysis for CAPAs needs to be well documented, showing actual tools used to support conclusions about the accepted root causes (or elimination of other potential root causes).

RCA is typically not done well because people leap to conclusions. They think they already know the cause. Or they think the formal RCA takes too long. They might think they already know how to fix the problem. Sometimes there is a fear of blame. Root cause analysis is NOT the same as who is to blame analysis! The blame game has no place in RCA.

For complex problems involving medical devices, a rigorous and methodical approach is warranted. Complicated quality issues require a more sophisticated approach to RCA. Per 21 CFR 820.100, "appropriate statistical methodology shall be employed, where necessary", to investigate the cause of nonconformities related to product, process, and the quality system.

Medical device companies sometimes have to deal with complex quality problems involving product in the field. At first, it is often unclear what the real problem is and how the medical device failure contributed. Like the Challenger disaster, there may be sketchy information initially. If customers have been seriously injured or even died, there is often intense pressure to understand the problem quickly. There is uncertainty about how many, where, and what happened. This is exactly when a rigorous and methodical failure investigation approach and RCA process are needed.

Complex quality problems often require the use of a team with a collection of the right skill sets to gather information and develop a problem statement. This includes establishing a team, analyzing the process and timeline, confirming facts, and defining what the problem is and is not. This may be followed by brainstorming techniques, RCA tools, and verifying assumptions and results.

A team approach is often used and brings in required SMEs (Subject Matter Experts). The CAPA owner or a project manager may be the team leader. Based on the initial assessment, team members might include process engineers or experts, product or design experts, material experts, suppliers, statisticians, quality systems experts, reliability engineers, regulatory, and medical experts. When conducting a failure investigation and determining the impact of a failure on the customer, it is important to consult qualified medical personnel. It may be necessary to add additional skill sets as the RCA proceeds.

The team should work together to develop the problem statement. This ensures that the correct problem is addressed. This forms the basis for the rest of the CAPA and it must be done well. Time spent on this stage prevents scope creep later on. The problem statement should indicate which feeder systems or sources of data are indicating an issue. Define when and where the problem occurred. How often is it occurring? Is it a trend? Is there a pattern? Provide adequate level of engineering detail. Warning: Do not state the cause!

Analyze the process. A tool such as a process flow diagram is useful to understand the process. It is also useful to highlight any recent changes to the process. Quality problems are often linked to changes in the process. These are things that may have occurred formally through the change control process. Or they could be changes like manufacturing equipment repair, new manufacturing operators, or new lots of raw materials or components. If a process, design, or application failure modes and effects analysis (FMEA), described later, were previously generated, they can provide useful insight. Review the FMEAs to see if the problem was previously identified and what risk level or occurrence was predicted. Or, perhaps this is a new failure mode that was never previously identified. In that case, the FMEAs must be updated.

Frame the process at a high level. Break it down to the main process steps and understand the inputs and outputs to the process. The SIPOC (Supplier, Inputs, Process, Outputs, Customer) tool described later is a very useful tool. Flow charts or more detailed process maps can also be used.

At this point it is possible to start identifying possible cause(s). Do not make the mistake of thinking there is only one root cause or an end cause. There may be a single cause, multiple causes, and multiple levels. Look for all levels and contributing causes. You can decide later which ones will be addressed. Tools like the 5 whys, a fishbone diagram, or logic tree can be very useful.

WARNING
___

Every failure involving product in the field (complaints, MDRs, or recalls) has at least two causes. There is a reason for the product defect itself. And there is a reason that it was not detected internally and escaped out to the field. Dig deeper and find the contributing cause(s).

Begin to collect data. This can be accomplished by interviewing SMEs, obtaining previous testing information, or reviewing complaint analysis. Determine when the problem started, if it is continuous or intermittent, when and where it does/does not exist. Conduct additional research, testing, or experimentation as applicable to understand the nature of the failure or nonconformity. During the Rogers Commission investigation of the Challenger disaster, renowned theoretical physicist, Richard Feynman, challenged NASA by asking them at what temperature they tested the O-rings. He famously conducted his own televised experiment. He put the O-rings in a glass of ice water and tested them with a c-clamp to demonstrate that the O-rings were less resilient at extremely cold temperatures and more subject to failure. Analyze the data to see if there are any correlations or patterns. Tools for repetitive events can include run charts, histograms, cluster diagrams, and scatter diagrams.

Create a **timeline** and tell the story in chronological order. This can be a timeline of the specific failure in use. Or it can be a timeline of events leading to the failure. For the Challenger, there is the timeline of the physical failure. Investigators played the launch tape frame by frame. At 0.5 seconds after launch, smoke first appeared. At 58 seconds, telemetry showed there was a loss of pressure in the right solid rocket booster. At 73 seconds, the shuttle broke apart in flames at 44,000 feet.

Include a timeline of when the problem was identified, what activities or changes occurred, when initial corrections were made, when containment activities were initiated, etc., all the way back to the triggering event. Every basic TV show detective knows how to use this simple but very useful concept. In the case of the Challenger, there is evidence that Morton-Thiokol knew about O-ring

Table 15.2 Confirmed Facts Table

ID	Confirmed Facts	Description	Supporting Data
1			
2			
3			

Table 15.3 Is/Is Not Example

Topic	Is	Is Not
What	What units, batches, or lots have the defect?	What units, batches, or lots do not have the defect?
	What is the specific failure mode?	What is the not the failure mode?
Where	Where is the failure occurring?	Where is the failure not occurring?
	Where are customers with failures located?	Where are customers without failures?
When	When does the failure occur?	When does the failure not occur?
Extent	How many units have the failure?	How many units do not have the failure?
	What is the trend?	What is not seen as a trend?

problems as far back as 1977. The events leading up to the launch indicate there were problems as early as 16 hours ahead of launch as the weather report predicted a launch temperature of 26°F.

Create a **list of confirmed facts** and evidence. Number the facts, provide a description, and supporting detail. This may seem like a very simplistic thing to do. But in the panic of a quality crisis, people easily lose sight of information or remember it in a way that supports preconceived notions. A confirmed facts table, with supporting information can keep information constant and accurate. It can also be a useful mechanism for communication. For the Challenger disaster, why did the right and not the left solid rocket booster fail? Confirmed facts were that the right solid rocket booster was in the shade at an estimated temperature of 32° F and the left was in the sun at an estimated temperature of 55° F (Table 15.2).

Create an **"Is/Is Not"** table to define the scope of the problem. Use confirmed facts. This tool may generate more facts to be added to the list of confirmed facts. It helps to determine the magnitude of the problem, not just the count. Ideally each "Is" should have a corresponding "Is Not." Remember to address both the creation and the escape of the problem. The tool may also raise questions about testing and confirmation (Table 15.3).

Brainstorm possible root causes. The fishbone diagram is an excellent brainstorming tool to encourage team participation. It is a graphic tool that helps to identify, sort, and display possible causes of a problem or quality characteristic. The **fishbone diagram** is also known as **Ishikawa Diagram, Cause and Effect Diagram, 6Ms (sometimes 5Ms and an E)**. This tool was made famous by Kaoru Ishikawa. The premise behind this tool is understanding and defining sources of variation in a process. These sources are typically identified as the **6Ms** (or sometimes **5Ms** and an **E**) for **M**aterial, **M**ethod, **M**achine, **M**easurement, **M**an (or human), and **M**other Nature (sometimes **E**nvironment). These sources of variation are displayed in a diagram that looks like a fishbone (hence the alternate name, fishbone diagram).

Pareto Analysis is a tool frequently used to identify the most prevalent sources of variation. It can be used to summarize and prioritize the sources of variation identified in a fishbone diagram. This tool is used to separate the "vital few" from the "trivial many" causes. These sources can be further investigated and addressed. The Pareto principle, also known as the "80/20 rule", says that 80% of the problem comes from 20% of the causes.

The **5-Whys Model** is a technique originally created by Japanese industrialist Sakichi Toyoda. It is a simple yet effective approach that even a young child can master! Simply keep asking why. Two-year old children are experts at using this tool. It is especially useful for management to use to understand an issue and the cause or contributing cause(s). It is not necessary to ask why exactly five times. But, you should keep asking why until you reach an appropriate level of understanding. It is an excellent tool for CAPA to get beyond direct root cause(s) to contributing causes. It can be used to get from physical causes of a failure all the way to understanding human cause. As in the Challenger example, it helped to move from seeing the O-rings as the physical failure mode to understanding the role of decision making and human causes. "Groupthink" is a psychological phenomenon that occurs within a group of people in which the desire for conformity results in irrational or dysfunctional decision making. The Rogers Commission ultimately flagged the serious flaw in the decision-making process leading up to the launch.

TIP

Engaged managers should use the 5-whys model often! Use it as part of affirmative responsibility to fully understand quality and compliance risk.

Fault Tree Analysis (FTA) is a top-down, deductive tool using Boolean logic to identify cause(s) of a failure. FTA is an excellent tool for low volume or sporadic events like a recall. It is particularly useful for determining the cause(s) of a known failure (such as a recall or MDR) because it shows layers of causes and their relationship to each other. Sometimes complex problems have multiple interacting causes that need to be understood. For example, a recall has at least two high-level causes: The reason a defect occurred *and* the reason it escaped your normal detection (acceptance testing) process and got into the field. Each of these high-level causes can be investigated further. This is an extremely useful tool for understanding complex product quality failures.

Conduct further testing to confirm hypotheses. **Hypothesis testing** consists of a null hypothesis and alternative hypothesis. Through a hypothesis test, a decision is made to reject the null hypotheses or not reject it. When a null hypothesis is rejected, there is an α risk of error.

A **contradiction matrix** can use known facts to either support or contradict a potential root cause. It is a useful tool to narrow down many potential root causes to a few to be studied. Any potential causes that contradicts known facts can be eliminated. Potential cause(s) that have no contradictions can proceed to further analysis.

TIP

This can be a very effective method of explaining to an FDA investigator why you chose not to react to a potential root cause.

Verify results and assumptions. Can the problem be reproduced? Does the team agree on root causes? Use the pareto method to determine how much of the problem each root cause accounts for. Use more tools such as DOE (design of experiments), ANOVA (analysis of variation), and hypothesis testing as necessary.

Verification is a very important step to ensure the correct root cause has been identified. Failure to do this step may cause the CAPA effectiveness check to fail. Or, even worse, it will not fail but the problem will resurface at a later point in time. Precious resources will have been wasted. Or patients may be impacted by RCA failure because quality problems were not completely resolved. Note: Validation and change control are addressed in the improvement part of the CAPA process.

All of the RCA steps above are not necessary for every CAPA. But they can give you an idea of effort, flow, and some common tools. More advanced methods may be needed for complex design or materials issues. And some issues may only require very simple tools such as the 5-whys model. A complex product quality issue involving product in the field may require more comprehensive use of tools.

TIP

Like everything else in the QMS, your CAPA, RCA, and failure investigation require good documentation! RCA conclusions should be documented in approved reports with supporting data, evidence of tools, and using appropriate statistical techniques.

Common problems with RCA include the following:

- One person comes up with the root cause on one's own.
- The description of what happened gets confused with the root causes of what happened.
- Too much urgency or a lack of resources results in incomplete analysis and inadequate problems resolution.
- There is no formal process for RCA.
- There is incorrect information or assumptions. Take time to confirm facts.
- There is failure to consider human factors as a root cause or contributing cause.
- Jumping to conclusions leads to inadequate depth of investigation.
- Blame or politics get in the way.

NOTE

The RCA above for the Challenger disaster is by no means complete. It is used for illustrative purposes only in this brief example of RCA. The Challenger disaster is often used as a case study for engineering safety, ethical decision making, and the dangers of groupthink.

An RCA process is necessary for every medical device manufacturer BEFORE you have a quality crisis. The toolbox should already have a variety of nonstatistical and valid statistical techniques. Not every tool is needed for every CAPA. It will depend on the type of problem and the risk involved. Create your RCA process and fill your toolbox before you need the tools. Ensure that you have personnel trained in RCA and appropriate statistical techniques. A methodical process and common lexicon are enablers of an effective and efficient QMS.

In summary, dig deep during RCA to find all root causes and contributing causes. Do not make the mistake of thinking it was *just* the O-rings. Key success factors include:

- Use of data and evidence
- Providing appropriate resources, training, coaching, and team engagement
- Use of a rigorous and methodical RCA process
- Providing an adequate toolbox with both nonstatistical tools and valid statistical techniques
- Not neglecting the human factor
- Weaving thorough RCA into a data-driven culture of quality and prevention.

Other tools useful for an efficient and effective QMS are outlined below. It is impossible to cover these tools and concepts in a complete manner in this book. Many of these topics can be an entire book by themselves. But, they provide a very useful set of tools for improving your QMS. I recommend that you look into these tools more deeply.

Plan Do Check Act (or PDCA) was made popular by the father of modern quality control, W. Edwards Deming. He also referred to it as the Shewhart cycle. PDCA is based on the scientific method of "hypothesis–experiment–evaluation." Deming later referred to the method as PDSA or "Plan, Do, Study, Act," because he felt that "check" emphasized inspection over analysis. Regardless of what you call it, PDCA is a useful model of critical thinking. PDCA is an iterative process repeated over and over. Each cycle extends knowledge.

Six Sigma (sometimes known as Process Excellence) is the most comprehensive problem solving and process improvement approach. It depends on good use of data and statistical tools. The methodology started at Motorola but was later used extensively at GE and others who credited it with saving them billions of dollars. I went through Six Sigma training in 1998 when it was just getting started. I was eventually certified as a black belt and eventually a master black belt. I've found six sigma concepts and tools to be invaluable in creating an efficient and effective Quality Management System. A lot of companies emulated the GE approach to Six Sigma, which involved charted projects and certifications (belt levels). In some cases, the GE approach did not fit well with company culture. Companies created too much bureaucracy around the chartering and certification steps. This gave Six Sigma a bad reputation at some companies. In some cases, high-level individuals in management were intimidated by the use of some statistical concepts. This created resistance to the use of Six Sigma at some companies.

Regardless of this resistance, the Six Sigma methodology and toolbox is still one of the best and most comprehensive approaches to process management, problem solving, and improvement. It is based on customer focus and statistical techniques to understand processes, sources of variation, and elimination of defects. That is all consistent with the expectations of an effective and efficient QMS. So, whether or not companies choose to use the GE concepts of chartered projects and certifications, they can and should take advantage of the rigorous, methodical approach to process improvement that is six sigma.

Six Sigma revolves around several key concepts that support an effective and efficient QMS:

- Customer focus and understanding what characteristics are critical to quality (CTQ).
- Process capability involves understanding and controlling sources of variation or KPIVs (key process input variables).
- Process control creates stable, predictable processes to improve customer satisfaction.

Six Sigma involves a road map known as DMAIC (Define, Measure, Analyze, Improve, and Control) that includes the following steps:

Define the Problem:
1. Define the problem, the defect, or nonconformity.
2. Define the benefit to customers and CTQs.
3. Determine defects and opportunities.
4. Determine the key output or "Y."
5. Select an appropriate team.
6. Define goals and timelines.
7. Identify team resources, sponsors, or management.
8. Charter the project.

Measure:
1. Validate the measurement system.
2. Establish process capability baseline.
3. Create a process map.
4. Identify inputs and outputs.
5. Implement containment activities.

Analyze:
1. Characterize the process distribution shape, center, and spread.
2. Develop improvement targets.
3. Identify sources of variation.

Improve:
1. Use hypothesis testing to determine critical inputs.
2. Establish operating limits.
3. Determine new process capability.

Control:
1. Monitor the process.
2. Document the process.
3. Implement process controls.
4. Verify long-term capability.
5. Obtain formal approval.
6. Check the effectiveness of actions.

Six Sigma methodology and tools can have a very valuable impact and application in an efficient and effective QMS:

- Design of new products
- Inspection, measuring, and test equipment
- Acceptance activities
- Creation of inspection control points, sampling plans, and test methods
- Process validation, optimization, and control
- Nonconforming product
- Corrective and preventive action.

Obviously, the techniques and tools of Six Sigma are well aligned with regulatory requirements and the process approach described so well in ISO 13485. Do not focus on using the method to drive belt certifications. Use the Six Sigma road map with a focus on desired improvements and getting results. Use the road map, tools, and lexicon with an intent to create a rigorous improvement approach. A rigorous and consistent approach facilitates a common language and expectations for data-driven problem solving. Six Sigma should be a part of your approach to an effective and efficient QMS.

Lean manufacturing concepts focus on systematic identification and minimization of sources of waste. Lean places an emphasis on the definition of value and what adds value from a stakeholder perspective. Use value stream mapping (VSM) techniques to carefully analyze current state and design future state QMS processes.

Change management is a discipline that guides how to prepare, equip, and support individuals and organizations to successfully adapt to change and drive successful outcomes. Good change management can accelerate the pace of improvement by identifying and removing barriers to change. It can help to understand who stakeholders are; shape a vision and common commitment to change; and make changes last. Change management is important for ensuring lasting improvements to the QMS.

C&E Matrix (cause and effect matrix) or prioritization matrix is a simple tool that can be used to narrow down a long list of suspected inputs to a more manageable list. It can also be used to prioritize improvement projects or activities. See Chapter 13, Translating Vision to Plans, for an example.

Cross-functional process map or swim lane map is a visual method used for process flow diagrams that distinguishes roles and responsibilities for the steps in a process. Swim lanes may be organized vertically or horizontally. The swim lane map differs from other flowcharts by grouping steps in lanes to distinguish responsibilities. It is useful for processes involving handoffs (Fig. 15.1).

Control Charts have been in use since the 1920s when Dr. Shewhart used them at Bell Laboratories. They are used to display data about a process over time. They are a graphical representation of a characteristic of a process showing plotted values versus a center line and one or two control limits. They include X bar and R charts, p charts, and more.

Control Plans are created to ensure processes are operated in a controlled manner. A simple control plan could be an extension of the control column of an FMEA. The concept of control plans is commonly used for manufacturing processes. But, any process can benefit from the idea of identifying and controlling sources of variation to improve process outputs. They can also be used to document and maintain improvements to prevent future process owners from undoing a deliberate improvement.

FMEA is Failure Mode and Effects Analysis. This is a qualitative tool that is useful in many applications and can help to methodically and rigorously analyze processes for what could go wrong and determine the severity, occurrence, and detectability. Severity, occurrence, and detectability are used to

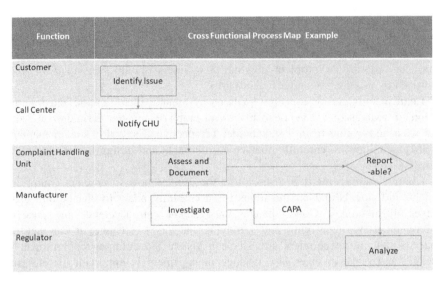

Function	Cross Functional Process Map Example
Customer	Identify Issue
Call Center	Notify CHU
Complaint Handling Unit	Assess and Document — Report-able?
Manufacturer	Investigate → CAPA
Regulator	Analyze

FIGURE 15.1

Cross-functional process map example.

determine a risk priority number (RPN). The medical device industry uses three types of FMEA:

- Use or applications FMEA: aFMEA
- Design FMEA: dFMEA
- Process FMEA: pFMEA.

FMEA is an excellent tool, not only for problem solving, but for risk management activities to proactively identify, evaluate, mitigate, and control risk. During RCA activities, the original FMEA(s), if existing, are very useful as a reference point.

FMECA (Failure Modes, Effects, and Criticality Analysis) goes a step beyond FMEA in that it also evaluates the criticality of the effect and actions that must be taken to counteract the effect.

A **heat map** is a graphical representation of data using colors to display risks. It is a very useful tool for visualizing and communicating risks to management.

Poka-Yoke is a mistake proofing concept that originated with Shigeo Shingo from Japan. The term is an Anglicized version of the Japanese words "poka" (inadvertent mistake) and "yoke" (prevent). Mistake proofing is achieved while the operation is happening. It relies on three steps:

1. Identify possible errors that could occur. At each step in the process, ask what human error or machine malfunction could occur.
2. Determine a way to detect the error.
3. Identify a specific action (warning, action, or control) to be taken when an error occurs.

Poka-Yoke and mistake proofing concepts are useful when documenting SOPs (standard operating procedures).

A **process capability index** is a statistical measure of the ability of a process to produce outputs within specification limits. Cp is a measurement of the allowable tolerance spread divided by the actual 6σ (six sigma) data spread. Cpk has a similar ratio but considers the shift of the mean relative to the central specification target. Process sigma level is a similar concept. These concepts allow process performance to be measured and monitored.

A **process flow diagram** defines the path of steps of work to produce or do something. Process mapping is very useful for understanding the steps of a process, decision points, and flow.

Requirements mapping is an effective method for making sure all regulatory requirements are identified and translated into specific SOPs and work instructions. It is helpful in making sure all requirements are present but not duplicated or even conflicting. It can be a useful guide for making changes and updates to procedures (Table 15.4).

Statistical Process Control (SPC) is the application of statistical techniques in the control of processes. It uses statistical concepts to understand the statistical

Table 15.4 Requirements Map

| Section | Requirement | | Document/ Procedure |
	21 CFR 820	ISO 13485:2016	
General	820.30 (a) General. Each manufacturer of any class III or class II device, and the class I devices (automated software; catheter, tracheobronchial suction; glove, surgeon; restraint, protective; radionuclide applicator; radionuclide teletherapy) shall establish and maintain procedures to control the design of the device in order to ensure that specified design requirements are met.	7.3.1 General The organization shall document procedures for design and development. During D&D planning, the organization shall document:	
D&D Planning	820.30 (b) Design and Development Planning. Each manufacturer shall establish and maintain plans that describe or reference the design and development activities and define responsibilities for implementation.	7.3.2 (d) Responsibilities and authorities for D&D	

(Continued)

Table 15.4 Requirements Map *Continued*

Section	Requirement 21 CFR 820	Requirement ISO 13485:2016	Document/ Procedure
D&D Planning	820.30 (b) Design and Development Planning. Each manufacturer shall establish and maintain plans that describe or reference the design and development activities and define responsibilities for implementation.	7.3.2 (a) D&D stages 7.3.2 (b) the review(s) needed at each D&D stage 7.3.2 (c) the verification, validation, and design transfer activities that are appropriate at each D&D stage 7.3.2 (e) the methods to ensure traceability of design and development outputs to design and development inputs.	
D&D Planning	820.30 (b) D&D Planning. Plans shall identify and describe the interfaces with different groups or activities that provide, or result in, input to the design and development process.	7.3.2 (d) Responsibilities for and authorities for D&D 7.3.2 (f) resources needed, including necessary competence of individuals	
D&D Planning	820.30 (b) D&D Planning. The plans shall be reviewed, updated, and approved as the design and development evolves.	7.3.2 As appropriate, D&D planning documents shall be maintained and updated as the D&D progresses.	

distribution of process performance outputs. There are three basic properties of a distribution: Shape, center, and spread. SPC requires understanding and controlling the sources of variation (common and special causes) to control and predict process performance.

SIPOC is an acronym for **S**uppliers (or source), **I**nputs, **P**rocess, **O**utputs, **C**ustomers. It is a high-level process map. Suppliers (or sources) are providers of process inputs and can be internal and external suppliers or sources. A supplier may be the previous process, a regulator, or an outside supplier. Inputs are the materials, resources, data, or information required to execute a process. The process is a structured set of activities that transform a set of inputs into outputs that provide value to outside customers, subsequent processes, or other stakeholders such as employees, shareholders, and regulators. This tool is useful to frame

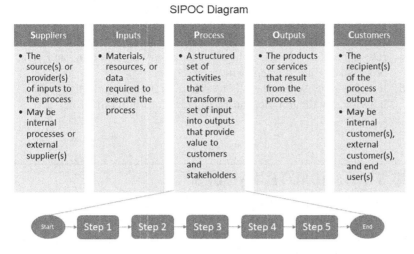

FIGURE 15.2

SIPOC diagram.

Table 15.5 RACI Model

	Project Leader	**Quality**	**R&D**	**Operations**
Activity 1	RA	C	C	I
Activity 2	C	I	RA	C
Activity 3	R	A	I	C

processes at a high level and to encourage thinking about connecting processes (Fig. 15.2).

RACI model (see Table 15.1), or responsibility assignment matrix, is a straightforward technique for identifying roles and responsibilities. It is useful for clarifying ownership to prevent gaps (ineffectiveness) or overlaps (inefficiencies) in the QMS. RACI is an abbreviation for:

- **Responsible**—the person who is assigned the work.
- **Accountable**—the person who makes the final decision and has ultimate ownership.
- **Consulted**—the person who must be consulted before a decision or action is taken.
- **Informed**—the person who must be informed that a decision or action has been taken.

Step 1. Identify processes or activities and list them on the left side column of the chart.

Step 2. Identify roles and list them along the top row of the chart.

Step 3. Complete all cells in the chart. Identify who has RACI for each process. Every process or project should have one and only one R.

Step 4. Resolve overlaps. An overlap occurs when multiple roles have an R. If resolution is not possible, then it is necessary to "zoom in" and break the process step into substeps with more clarity.

Step 5. Resolve gaps. Where there is no role identified for an activity, the individual with authority must assign responsibility to a role.

VALID STATISTICAL TECHNIQUES

This one is not optional. The quality system regulation explicitly requires, in 21 CFR 820.250, that medical device companies "establish and maintain procedures for identifying valid statistical techniques required for establishing, controlling, and verifying the acceptability of process capability and product characteristics." Further, "Sampling plans, when used, shall be written and based on valid statistical rationale. Each manufacturer shall establish and maintain procedures to ensure that sampling methods are adequate for their intended use and to ensure that when changes occur the sampling plans are reviewed."

Valid statistical techniques is a very large topic that cannot be covered in this book. It is important to have personnel with demonstrated education and skill on this large topic. A degree in statistics, applied statistics, or biostatistics is appropriate for personnel setting policies and procedures for statistical techniques. A degree in engineering, certification as a black belt or master black belt in Six Sigma, or ASQ (American Society for Quality) certification in quality engineering is appropriate for making decisions about sampling plans, acceptance testing, control charts, trending, etc.

Value stream mapping is a useful tool for analyzing processes in terms of value added and may be used to improve process efficiency. A value stream is the steps and all the actions in a process. A value stream implies looking at the whole process to identify value-added and nonvalue-added steps. VSM is a lean management tool for analyzing current state and designing a future state.

FDA inspection preparedness

16

Medical device manufacturers should always be ready for a regulatory inspection. Inspection preparedness is not just a quick cleanup of the manufacturing floor when you know that a regulator or notified body is coming. Inspection preparedness requires that you consistently follow and document the basics of a quality management system (QMS) on an ongoing basis:

Say what you do, do what you say, and document it.

It also requires planning, preparation, and practice. The Food and Drug Administration (FDA) and International Medical Device Regulators Forum (IMDRF) have several documents that should be reviewed and understood to develop an effective inspection preparedness program:

- *Guidelines for Regulatory Auditing of Quality Management Systems of Medical Device Manufacturers*, Parts 1−5 from GHTF Study Group 4;
- *Guide to Inspections of Quality Systems or Quality System Inspection Technique* (most often referred to as *QSIT*);
- *FDA Investigations Operations Manual.*

The goal of investigators during inspections is to verify:

- The company is complying with applicable regulations.
- The QMS is effective at preventing the occurrence of product issues that could affect the public health.
- Processes are established and maintained to comply with regulations.

There are different types of FDA inspections. The investigator's focus depends on the type of inspection:

- Periodic inspection using the Quality System Inspection Technique (QSIT)
- Pre- and post-PMA (premarket approval process) inspections
- Special inspections:
  - For cause (e.g., a class I recall or increase in Medical Device Reports (MDRs) show quality issues and increasing risk to the public health)
  - Risk-based work plan
- Foreign inspections.

Medical Device Quality Management Systems. DOI: https://doi.org/10.1016/B978-0-12-814221-9.00016-6

A current focus of regulators is the Medical Device Single Audit Program (MDSAP). MDSAP was initiated in 2012 at the International Medical Device Regulators Forum (IMDRF) inaugural meeting. IMDRF members are using MDSAP to leverage resources to manage an efficient and effective single audit program. The pilot program was initially greeted with some skepticism by manufacturers. But it is now seeing more common usage and many companies have found it to be positive. Some countries (Brazil) utilize the outcomes of MDSAP to constitute an important part of pre- and postmarket procedures. Health Canada will require MDSAP in 2019 for Canadian Medical Device Conformity Assessment System (CMDCAS) certification. The United States has indicated that it will accept MDSAP reports as a substitute for routine inspections. Inspections that are considered "for cause" or "follow-up" will not use MDSAP.

The QSIT approach used by the FDA covers four key areas:

- Management responsibility
- CAPA (and satellite processes, medical device reporting, corrections and removals, and device tracking)
- Design controls
- Production and process controls (and the satellite process, sterilization process controls).

The FDA QSIT approach includes three levels:

- Level 1—Abbreviated inspection only covers two subsystems (CAPA plus one other subsystem) and is, therefore sometimes referred to as "CAPA plus 1." It is conducted after a firm has had a level 2 inspection and was found in compliance with the regulations.
- Level 2—Full (comprehensive), covers all four subsystems. It is conducted when a firm has never had an inspection and every 6 years thereafter.
- Level 3—Uses QSIT as guidance but is more comprehensive. Uses when the previous inspection was official action indicated (OAI) or as directed by CDRH/Office of Compliance.

Pre-PMA inspections are done before premarket approval (PMA) and include:

- Design verification and validation with a focus on new features in product specifications (i.e., how does this product differ from previous inspections)
- Process qualification and validation with a focus on new processes first
- Clinical or outside U.S. (if in other markets) product performance

Post-PMA inspections are done after PMA and include:

- Design changes made since release
- Process changes made since release
- Product performance including returned products, complaints, and associated MDRs
- CAPAs for that specific product.

Special inspections may be done "for cause." They may also be based on the FDA's risk-based work plan. "For-cause" inspections include:

- Follow-up on observations from previous inspections (approximately 6–12 months after a warning letter)
- Recalls or market withdrawals
- Increasing adverse event reports, trends, or new failure modes
- Suspicion of fraud. Sometimes suspicions are aroused by anonymous tips or even whistle-blowers from inside the medical device company.

The FDA's risk-based work plan is developed to focus limited FDA resources on high-risk companies or issues. It uses science-based risk management to select and prioritize sites to be inspected. These inspections are initiated based on the request of CDRH rather than district offices. At the writing of this book, we do not know exactly how this will be impacted by use of the proposed voluntary submission of product quality metrics. But, expect some changes.

Foreign inspections are conducted by the FDA outside of the United States and Puerto Rico. Foreign inspections are QSIT level 2. Because a US Federal Agency is conducting business in another country, they have to notify the other country and the manufacturer as well. Therefore, foreign manufacturers always have the advantage of receiving advance notice of foreign inspections. Additionally, investigators typically conduct multiple inspections when traveling abroad with strict timelines and travel plans resulting in shorter inspections of predetermined length. For these reasons, foreign inspections are sometimes considered a little bit easier.

NOTE

The FDA uses the term "inspection," not audit, for their activities. FDA personnel have varying job descriptions but those who conduct inspections are usually referred to as "investigators." Some other titles you may encounter are consumer safety officer, national expert in design controls, commissioned corps officer, investigative engineer, microbiologist, or other specialties.

NOTE

Section 704 of the FD&C Act makes it a criminal act to refuse a lawful FDA inspection.

Investigators must show their credentials and issue a Form 482, Notice of Inspection, when arriving at a facility. If samples are taken, investigators must leave a receipt, Form 484. FDA investigators are trained to determine if violations of law exist and obtain voluntary compliance if possible. If voluntary compliance is not obtained they will document facts needed to support regulatory action, including litigation, if necessary.

The usual sequence of events for an FDA inspection is:

- Arrival
- Issuance of Form 482, credentials, and introduction
- Generally, a facility tour
- Conduct inspection in accordance with guidance documents
- Sometimes, collection of samples
- Issuance of a Form 483, if necessary, at the end of the inspection.

FDA inspection techniques include inquiry, observation, and contemporaneous, sequential note taking. This is similar to internal auditing. What is very different from internal auditing is that FDA inspections include documenting evidence of individual responsibility. The FDA trains its investigators on why and how this must be done. The *FDA Investigations Operations Manual*, section 5.3.6, says that "The identification of those responsible for violations is a critical part of the inspection, and as important as determining the violations themselves. Responsibility must be determined to identify those persons to hold accountable for violations, and those with whom the agency must deal with to seek lasting corrections."

In order to support further regulatory actions, FDA investigators are trained to establish JIVR (Jurisdiction, Interstate Commerce, Violation, and Responsibility). To establish responsibility, investigators will collect full names, titles, business cards, organization charts, verbal statements, or look for company publications, letters, memos, etc. Investigators will attempt to gather information about who knew or should have known about issues and who had a duty and power to correct things. Investigators will attempt to connect these to specific violations. They may ask for shipping and distribution records to document interstate commerce.

FDA investigators are skilled at interviewing techniques. To resolve uncertainties, investigators will compare:

1. Answers peer to peer
2. Answers of subordinates to their supervisors and vice versa
3. Answers from interviewees to what the records show
4. Answers from interviewees to what policies and procedures require
5. Answers from interviewees to what inspection of work areas, equipment, and product demonstrates.

Prepare, prepare, prepare. Every medical device company should have an inspection preparedness process! The management representative must ensure plans are in place, roles are defined, preparations are made, and practice sessions are conducted. The management representative must ensure a cooperative and accurate inspection.

Like any other effective process, the inspection preparedness process should have metrics for monitoring and controlling performance. Metrics for inspection preparedness define your ability to successfully manage the logistics and dynamics of an external inspection. Consider your skills in managing an external regulatory inspection:

- Front and backroom preparedness
- Roles and responsibilities defined
- Percentage of personnel that have little, medium, or extensive experience
- Percentage of personnel that are trained on inspection preparedness
- Percentage of items from the inspection preparedness checklist that are completed
- Does the site have adequate resources, trained personnel, opening presentations, process diagrams?
- Have mock inspections been completed?
- Are records reasonably accessible? Have a plan for getting documents including those stored offsite (e.g., Iron Mountain). Conduct a test.

Determine key inspection roles and responsibilities such as:

- Inspection leader or coordinator
- Subject matter experts (SMEs)
- Backroom leader
- Scribe.

The inspection leader, a key role, coordinates activities and communications to ensure the inspection goes smoothly. They are the key contact point with the investigator. These individuals spend the most time in the front room and guide the discussion with the FDA investigator. They provide background and context for other individuals and SMEs that may come into the front room. The inspection leader may be the management representative or a delegate. The inspection leader helps investigators understand processes and formulate requests. They must identify appropriate SMEs and other responders. The inspection leader:

- Ensures requests are fulfilled in a timely manner
- Facilitates clear communication between investigators, SMEs, and other participants
- Ensures that the site, management, and other stakeholders are aware of inspection activities, status, and priorities
- Monitors front and backroom effectiveness
- Reviews issues presented by the backroom
- Strategizes with site participants on critical items and responses.

A scribe should also be in the front room to take notes in an electronic format that can be transmitted to the backroom real-time. Accurate notes and information are very helpful for letting the backroom know what is being discussed and what may be coming up next. (Hint: Although the role of the scribe in not critical, it is a terrific opportunity for personnel to learn and understand managing an FDA inspection. Use it to develop essential inspection personnel.)

SMEs are also critical to inspection success. Investigators want to speak to personnel that have first-hand knowledge of processes or activities; can explain decisions made; and can confirm that processes are defined, understood, and implemented. The inspection leader should select SMEs that can answer questions correctly; provide complete and accurate explanations; present information clearly and concisely; and know when to speak up and when to remain quiet. They should be able to communicate at a high level and go deeper when necessary and understand how their activities fit within the overall quality system. They must be able to handle the pressure, be willing to take direction from the front and backroom, and take feedback without getting defensive. Above all, they must be able to instill confidence.

Yes, that is a tall order. Trusted SMEs are worth their weight in gold during an FDA inspection. Successful inspection leaders must know their organizations well to find individuals with those skills. Or, they take active preparations to train and cultivate SMEs. Two areas where SMEs are especially important are the complaint handling and CAPA processes. Your complaint handling and CAPA leaders need to be able to handle regulatory inspections well!

A well-managed backroom is also important to a successful inspection. The backroom may be working behind the scenes, but they play a key role. An experienced backroom leader monitors the scribe notes from the front room and has an idea where the conversation is heading. They have an awareness of sensitive issues and potential problems. They should anticipate needs and be proactive in getting information ready. They should be able to handle the pressure, work in a rapidly changing environment, handle multiple priorities, pay attention to details, communicate clearly and concisely, work quickly and accurately, set realistic timelines for delivery of requests, and accurately fulfill requests from the front room.

Backroom personnel can make or break an inspection. Sometimes multiple people are needed to fulfill requests, check materials, and prepare SMEs. If you have more than one investigator, you will need additional backroom personnel.

Most medical device companies use the front room/backroom approach for managing an inspection. Sometimes the backroom is known as the "war room" but that creates an image of confrontation and fighting. A backroom mentality with expectations of competence, correctness, calmness, and confidence is a more useful image.

During an inspection, investigators will meet in the front room with the inspection leader and appropriate site participants. Additional site participants will be in the backroom. The front- and backrooms serve very different purposes. The front room must:

- Facilitate the inspection
- Facilitate communications amongst investigators, SMEs, and the backroom
- Provide a sense of calm and confidence.

- The inspection leader must monitor the overall flow and content of information going to the investigator(s). Multiple investigators will require additional front room leaders to monitor the flow of information. Investigators will dig into any inconsistencies in information.

The backroom must:

- Provide the front room with accurate documents and records in a timely manner;
- Prepare and coach SMEs and other personnel before they go into the front room.

Records are a critical element during an inspection. Investigators' requests for documentation should be fulfilled in a timely and accurate manner. Make sure you have an efficient, effective method for managing records! If you are late in providing one requested record, it will most likely not be a problem. If you are consistently late, it will create a negative impression.

Inspection preparedness plans should define the front and backroom location and setup. The front and backrooms should be clean and orderly before an inspection starts. The front room should:

- Ideally be near the public entrance so you do not have to lead an investigator through long hallways and past employee desks
- Contain equipment necessary for an inspection (table and comfortable chairs, whiteboards, or projectors if requested)
- Have enough power outlets for multiple computers, projectors, etc.
- Have Internet access for scribe notes and communications
- Be connected to the backroom via scribe notes and/or instant messaging (or other communication method).

The backroom should:

- Be near, but not in direct view of the front room. It should be near enough to facilitate bringing records to the front room in a timely manner. But, you do not want the investigator to see (or hear) the backroom and a flurry of nervous activity. Investigators all know that backrooms are used, but you do not need to make it obvious or show signs of stress.
- Contain equipment necessary for fulfilling requests:
  - Dedicated color printers
  - Projectors for viewing front room scribe notes
  - Supplies such as paper, printer ink cartridges, staplers, folders, and other organizing materials
  - Stamps to mark documents as "Company Confidential" or "Copy" (or print with watermarks)
  - Many power outlets and Internet connections. You will need these for many laptops, printers, and projectors.

It is not a question of if, but when, an inspection will occur. Therefore, proper planning and preparation are required! The management representatives, or designated inspection leaders, are vital to managing external regulatory inspections. They must plan and prepare for inspections, oversee the inspection process, and follow up after an inspection. Preparations must include the following:

1. Site inspection preparedness program.
   a. Review your site's history of inspections including any previous Form 483s, warning letters, and establishment inspection reports (EIRs). Review any responses and commitments made.

> **TIP**
>
> Past EIRs are a valuable source of information. Be sure to read them carefully. Every FDA investigator uses the last EIR as a starting point for their inspection. Minor concerns mentioned in a past EIR can be a real Form 483 observation the next time. Always follow up with the FDA if you have not received an EIR within a few months of an inspection.

> **TIP**
>
> Large companies with multiple sites can benefit from sharing inspection results, lessons learned, and EIRs with each other. When the FDA sees similar issues in multiple sites, they may conclude that your company has systemic issues. Create a formal mechanism for sharing information.

   b. Be aware of any significant quality issues (class I recall, several recalls, increase in MDRs) or employee (i.e., whistle-blower) concerns that might trigger a "for-cause" investigation.

> **TIP**
>
> Watch your contract sterilizers and contract manufacturers for any regulatory issues with possible health hazard consequences. That may trigger an inspection at your facility.

   c. Define key roles and responsibilities for inspections including inspection leader, backroom leader, scribe, and common (e.g., complaint and CAPA) SMEs.
   d. Create a contact list for key personnel. Include phone numbers and mobile phone numbers for key personnel. Critical personnel (management representative, inspection coordinator, backroom leader, etc.) should all have two layers of delegates or backup. Note: A management representative is a key organizational role and should always assign a delegate during planned absences.
   e. Create a training plan and training materials for all roles.

f. Develop a communications plan. Identify who needs to be informed at the beginning, during, and after an inspection. Make sure the communications plan addresses the needs of different types of inspections (e.g., a pre-PMA inspection will involve different personnel than a for-cause inspection).

g. Plan for a method of documenting, fulfilling, and tracking requests. Prepare a request-tracking tool that identifies when a request was made, and when and how it was fulfilled.

  i. Determine necessary skills and authorizations for pulling electronic and/or manual documents and running reports. Make sure you have these taken care of ahead of time. If applicable, make sure you have agreements with outside storage services (e.g., Iron Mountain) to locate and deliver needed paper documentation promptly.

  ii. Make sure you have stamps or watermarks that identify documents as company confidential.

  iii. Add a checking step for all requested documents. Make sure the backroom or SME reviews all documents thoroughly for content, issues, or inconsistencies prior to taking them in the front room. Any issues identified should be discussed with the SME and inspection leader ahead of time.

  iv. Review all documents and records for good documentation practices (GDP) violations before documents go into the front room! These are rookie mistakes that immediately create a very bad impression with investigator(s). Common GDP errors include:

   1. Missing signature and date
   2. Write overs, scribbles, white out, or other unacceptable changes
   3. Nonuniform date and signatures
   4. Date entries that do not match with batch record timelines
   5. Blanks on production records
   6. Prerecording of data
   7. Incomplete references
   8. Illegible writing
   9. Too many corrections
   10. Sticky notes in documents
   11. Pencil instead of permanent ink.

  v. Keep identical duplicates of all documents provided to the FDA.

h. Establish, in advance, a policy for controlling use of cameras, and video or audio recording devices (also firearms or other restricted items) by external regulators and other visitors. Most companies have a strict policy that no photos or recordings are allowed on-site. Consult your legal counsel ahead of time to define your approach.

i. Establish a policy or practice for allowing investigators to see IT systems (CAPA, complaints, calibration, etc.) in real-time demonstrations. Also have a policy and expectation for providing investigators Excel spreadsheets on a flash drive or other storage media to be taken and analyzed by the FDA. Keep duplicate copies. Be prepared for investigators to ask for custom queries from IT systems.

j. Plan and schedule practice sessions or mock inspections.

k. Establish inspection norms and rules such as
   i. Using the contact list
   ii. Marking documents as confidential
   iii. Checking documents before taking them into the front room
   iv. Logging all requests
   v. Controlling the flow of documents
   vi. Attending daily debrief meetings
   vii. Working accurately, with a sense of purpose, while maintaining a calm environment.

l. Create a written plan that is reviewed with management. Keep it readily available. The plan should include an annual review and update. Key issues such as a class I recall or class action litigation should prompt an additional review and update.

m. Create inspection preparedness checklists and assign owners.

2. Prepare commonly requested documents such as:

a. A site overview presentation giving basic information about the site including size, number of people, types of products, relationship to other sites, key products and processes.

b. Organizational charts should be kept up to date on an ongoing basis. Be sure to identify the management representative and their relationship to management with executive responsibility.

c. Some process owners may prepare by creating a brief process map or flowchart to describe an important or complex process such as complaint handling or CAPA.

3. Prepare the infrastructure:

a. Identify equipment needs, including backups. Maintain a list of needed items (staplers, folders, stamps, paper, pens, highlighters, paperclips, binders, flash drives, spare ink cartridges for the printer, etc. Do not forget tissues just in case!). Consider having a dedicated box of supplies and printers to be used only for inspections. Check it periodically to make sure it is secure, current, and has not been raided. You do not want to have to run around at the last minute trying to gather supplies or get printer ink refills. Make sure you have a dedicated high-speed, color printer for inspections. The printer should be dedicated to inspection use only. Do not let your information flow to the FDA get bogged down by a slow, old-fashioned printer or other personnel printing big, noninspection documents between critical inspection documents! It is helpful to have a printer in the backroom, so inspection personnel are not running back and forth to the printer down the hall.

b. Make sure front and backrooms have needed power and Internet connections. Plan for this ahead of time so that you are not looking for extension cords at the last minute.

c. Review any IT system changes, planned maintenance, or outages.

d. Always notify the IT organization of inspections. You do not want your document management system, compliant handling system, or CAPA system to be down for routine maintenance during an inspection. Additionally, you may be asked to show how these critical systems were validated.

4. Prepare the people:

a. Identify all personnel that may be involved in inspections. Remember personnel that may have a minor role but may still be impacted (e.g., during a plant tour, an FDA investigator may walk up to a random manufacturing operator and ask them what the quality policy is).

b. Train and coach all participants. Critical roles should all have initial and refresher training. If it has been a while since the last inspection, do refresher training. SMEs that have little practice with inspections may receive refresher training and coaching from the backroom as needed.

c. Prepare people for various FDA inspection techniques such as:

    i. Pausing during interviewing as though they are waiting for more information from the interviewee. This makes the interviewee uncomfortable and likely to ramble on and provide too much information.

    ii. Arguing, pestering, or making snide remarks to frustrate personnel prompting them to blurt out additional information.

    iii. Use of leading questions, hypothetical questions, open-ended questions, and confrontational questions.

    iv. Questioning if they must follow clean room or gowning practices. Yes, they do.

d. Make sure you have skilled personnel in the backroom who can quickly and accurately locate electronic or hard-copy documents. Make sure they have appropriate authorization to access electronic files.

e. Conduct practice sessions/mock inspections if you have inexperienced personnel. Mock inspections are very useful for preparing an organization for inspections.

WARNING

Do not make the mistake of confusing an internal audit with a mock inspection. Audits and mock inspections have completely different purposes. An audit is intended to self-identify nonconformities and opportunities for improvement. An audit is most useful when approached with openness and transparency. By contrast, a mock audit should prepare an organization for an external regulatory inspection, where information must be shared in the most favorable manner. A mock inspection should simulate the pressures and stresses of a real FDA inspection.

    f. Ensure that training records for all company personnel are up to date and documented! This is especially true for key individuals such as the management representative, inspection leader, and SMEs that interact with the investigator in the front room. The investigator may request the training records of these individuals or someone they select randomly during the plant tour.

    g. Prepare security or reception: Your receptionist or security personnel are the most likely first contact for the FDA. Make sure they have a current contact list and know who to call. They should have a designated place to have FDA investigators wait until the management representative, inspection leader, or delegates arrive.

TIP

Make sure you have a quiet, confidential area for investigators to wait. Do not send them unescorted to wait in the cafeteria or a management office where they might see or overhear confidential information.

---

5. Prepare the facility:

    a. Have a plan for where to locate the front and backrooms. An inspection is inevitable, so do not wait until the last minute.

    b. Consider what bathroom facilities will be available to investigators. You do not want investigators to overhear confidential bathroom conversations or comments like "I'm so relieved the investigator didn't look at the CAPA XXX (or complaint trends for product YYY)." If you have a dedicated, quiet bathroom for investigators to use then certainly you can let the investigator have privacy. Otherwise, you need to have a discrete way (e.g., scribe notes indicate to backroom that a bio break is coming up) to inform the backroom to clear the path and bathroom. Make sure there are no distractions, issues, or unintended communications.

6. Prepare management:

    a. Ensure management is trained and understands the role of management with executive responsibility. Let the site manager know that they can meet the investigator and express their cooperation and commitment to sharing information accurately. Note: Sometimes top site management wants to attend the entire inspection and that is a good practice to facilitate active management responsibility and a culture of quality and compliance.

7. Practice. Practice. Practice. During an inspection, an organization needs to respond promptly and correctly. Practice ahead of time to make sure the following actions are well executed:

    a. The method for scribe notes and sharing communications with the backroom works consistently and efficiently. The backroom needs to see the running commentary from the front room real time to understand where the conversation may be heading. Many companies display scribe notes via a projector in the backroom. Other organizations (sites or functions in different geographic locations) that need to be informed may have controlled access to scribe notes.

b. The request fulfillment process works consistently and correctly.

c. Large multisite organizations need to be able to interact with other facilities. Identify these sites (e.g., the complaint handling unit (CHU)) and make sure you have them in your communications plan. Make sure you have an efficient method for communication and for potentially sending large files.

d. Consider creating layouts for front and backrooms, including equipment and people. Some organizations go as far as creating seating arrangements for key personnel. For example, in the front room make sure the inspection leader (and not the investigator) has visibility to the door. Other companies even have backroom layouts, so they have room for all necessary personnel and to facilitate communications (e.g., put complaint handling, adverse event reporting, and recall handling personnel together at a table or "pod" in a for-cause inspection). Plan for where to put equipment such as the high-speed color printer, or screen for projecting scribe notes.

Be prepared for anything during an FDA inspection. Internal personnel will be stressed and nervous and may act out of character. Inexperienced personnel may choke or even break down in tears. Some may display anger or frustration (Note: Sometimes investigators deliberately try to incite this.) and then blurt out inappropriate information. Others will surprise you by rising to the challenge and blossoming under pressure. But, proper planning and preparation in advance will ease some of the pressure. Make note of individual performance as you prepare for your next inspection.

Some FDA investigators are more experienced than others. Some are focused on the big picture and put an emphasis on quality. Others will be nitpicky and will not stop until they find some small compliance matter to write a Form 483 observation for. Some will spend significant amounts of time reading procedures and others will spend more time asking probing questions. Some will take documents away with them and come back the next day with more questions. Watch for interactions when multiple investigators are involved. Sometimes a more junior investigator will accompany a senior investigator. Some companies will have repeat investigators year after year, and others will experience someone new each time. Try to find out as much as possible ahead of time regarding the particular investigator that arrives at your door. Everyone will be different. And issues that are completely unrelated to the inspection may have a significant impact.

CASE STUDY 1

In one FDA inspection, a facility of approximately 30 individuals had to host three FDA investigators for a period of 6 weeks. This tiny facility only had one conference room. The three FDA investigators were often in disagreement with each other creating unique stresses we had never experienced. We had to figure out how to separate investigators and use office space (generally a no-no) to get them out of the tense and overly crowded conference room. Plant personnel were stressed, worked long hours and weekends, and had to put routine work on the back burner for 6 painfully long weeks. Personnel from other facilities (including me) were called in to support this site. The only bathroom in the office area was right next to the backroom, requiring extra care to prevent investigators from overhearing backroom conversations. The facility was cramped and crowded. It was a grueling experience for all involved. Even though the site was issued a Form 483, it was later successfully challenged due to the inconsistencies and disagreements between investigators. This was a highly unusual inspection, but I use it as an example that you must be prepared for anything.

CASE STUDY 2

I had to provide inspection support to a manufacturing facility in Juarez, Mexico. The inspection proceeded routinely for several days at the manufacturing facility. Each day after the inspection the investigator and support staff had to cross the border to get back to the United States and spend the night in a hotel in El Paso, TX. This was shortly after 9/11, and border controls were very tight, sometimes taking 2−3 hours to cross the border. We became aware that President George W. Bush was to be visiting El Paso on the last day of the inspection. We knew that border security would escalate to another level with even more severe delays. In this case, we were able to work with the FDA investigator to finish the last day of the inspection in the hotel in El Paso, very near to the airport. At lunch, we all went outside to watch the presidential motorcade go by. After the inspection successfully wrapped up, the cell phones starting ringing as another facility was notified of an FDA inspection and we were called in for support. And we were off to the airport again.

Create your own inspection preparedness checklists. You can use the checklist below as a starting point. But, adapt it for your own needs and product types. Create detailed checklists by process area (a best practice for process owners). You can even add a column for a responsible person if necessary for action or follow-up.

WARNING

This checklist is NOT intended to be used as a replacement for a fully functioning and effective QMS. But, it can provide some last-minute preparation guidance. Best practice would be to have individual process owners create their own inspection preparedness checklists and go through them periodically.

NOTE

The checklists contain items other than quality system requirements. It is important to also pay attention to other regulatory or safety requirements. Something as simple as a blocked eye-wash station or fire extinguisher on the manufacturing floor may capture the investigator's eye and create an impression of carelessness. And if a facility is seen as careless about safety, then what about quality? Pay attention to the basics. Order and cleanliness are important! Delta Airlines got a warning letter from the FDA in 2011 that mentioned "rodent excreta pellets (too numerous to count) and rodent urine stains" in ceiling panels and near food preparation areas. Yes, really.

Always make sure things are neat, clean, tidy, organized, and in control. Segregation and control of materials is always critical. Clear labeling, signage, and information throughout the facility creates an impression of thoroughness and control.

When an Inspection Is Announced	Yes	No	Follow-Up Required
Review and refresh the contact list. Create email distribution lists as appropriate for daily debriefs.			
Review the status of any previous responses, CAPAs, or commitments made to the FDA. Be prepared to discuss these commitments.			
Refresh overview presentations if necessary.			
If requested by the FDA, provide advance copies of the Quality Policy, Quality Manual, and requested high-level procedures. These will be returned during the inspection.			
Review key performance metrics (MDRs, complaints, CAPAs, recalls, etc.). Identify trends or issues that might be questioned.			
Notify key personnel per your communication plan.			
Notify and make sure site management/leadership is prepared to greet the investigator, demonstrate cooperation, and establish their commitment to supply needed information.			
Notify the entire site that an inspection is in progress. Notify them that the FDA may be taking a tour through the facility. Remind them of the importance of the Quality Policy, their own training documents, and having written procedures available at point of use.			

(Continued)

Continued

When an Inspection Is Announced	Yes	No	Follow-Up Required
Reserve the front and back rooms. Block any other meetings from occurring in or directly near the front room (keep noise and disruptions down).			
Give refresher training, as necessary, for inspection personnel.			
Ensure the back room has training materials to prepare any SMEs necessary at the last minute.			
Set up the front room. Ensure the front room is clean, organized, and has appropriate equipment. Make sure supplies, power connections, internet connections, etc. are ready.			
Set up the back room. Ensure the back room is clean, organized, and has appropriate equipment. Make sure supplies, power connections, internet connections, etc. are ready. Make sure the back room has a dedicated printer used ONLY for the inspection. Make sure a projector (and spare projector or spare bulbs) are available to review scribe notes.			
Post "do not disturb" signs on the front and backroom doors indicating that an inspection is in progress.			

Management Responsibility Checklist

Ensure organizational charts are current and correct. The management representative and highest level management must be identified. Print copies.			
Be prepared to discuss the Quality Policy and Objectives. Be prepared to show the Quality Manual. Be prepared to discuss quality planning.			
Be prepared to show the procedure for management review.			
Review the management review schedule and be prepared to provide documentation that management review meetings have been conducted as required. It is not necessary to show actual management review materials.			
Be prepared to the show the procedure for internal audit.			
Review the internal audit schedule and be prepared to demonstrate that audits have been conducted as required. It is not necessary to show actual audit reports.			
Be prepared to identify the management representative, delegation letter, and organization chart showing relationship to management with executive responsibility.			

(Continued)

Continued

When an Inspection Is Announced	Yes	No	Follow-Up Required
Be prepared to show that the quality policy and objectives have been implemented. Show the Quality Policy is visible on posters, communications, computer login screens, etc. Be prepared to provide training records demonstrating personnel have been trained on the quality policy.			
Be prepared to discuss quality plans. Discuss how the Quality Policy, the Quality Manual, DMRs (Device Master Record), and procedures work together to ensure product quality.			

Corrective and Preventive Action Checklist

	Yes	No	Follow-Up Required
Prepare an overview or flowchart of the process. Provide procedures showing the requirements of the regulation have been met.			
Ensure that software validation plans, reports, and an SME are available for the IT system, spreadsheets, or any databases used.			
Establish the criteria and format for any system-generated reports presented to investigators.			
Establish a consistent format for any CAPA files provided to the investigator.			
Review sources of data for on time, accurate reports. Ensure process owners can explain the reports, trends, outliers, and any out of control conditions.			
Identify and review CAPAs likely to be reviewed by investigators depending on the scope and reason for an inspection. Note: 2 years of data is often requested by an investigator as that is a typical interval between inspections.			
Identify and review high-risk CAPAs likely to be reviewed by investigators. Examples: CAPAs for recalls, MDRs, systemic issues.			

Complaint Handling/MDR

	Yes	No	Follow-Up Required
Prepare an overview or flowchart of the process. Provide procedures showing the requirements of the regulation have been met.			
Ensure MDR event files are prominently identified and easy to access.			
Ensure that software validation plans, reports, and an SME are available to discuss the IT system, spreadsheets, or any databases used.			
Evaluate the status and on-time reporting of all MDRs.			
Evaluate all decisions and rationale to report or not report complaints as an MDR. Note: Decisions not to report must be documented in the event file.			

(Continued)

Continued

When an Inspection Is Announced	Yes	No	Follow-Up Required
Establish the criteria and consistent format for any system-generated reports provided to investigator.			
Establish a consistent format to provide any complaint or MDR files to the investigator.			
Plant Tour—Returned Products Lab			
Review and follow the procedures for restricted access areas.			
Ensure that training and certifications are current for all personnel.			
Ensure that product is stored/maintained/controlled per requirements.			
Ensure that nonconforming product is dispositioned per requirements.			
Ensure that equipment maintenance and calibrations are current and acceptable.			
Review biohazard requirements and status.			
All signage (biohazard, etc.) must be in place and legible.			
Plant Tour—Manufacturing/Production Areas			
Review applicable gowning and entry requirements. Ensure FDA investigators comply with procedures (Note: They might test you to see if you will enforce requirements).			
Review ESD (electrostatic discharge) or antistatic requirements.			
Ensure safety glasses, gowning, and other requirements are in place and followed.			
Ensure fire extinguishers, emergency exits, eyewash stations, etc. are all clear and accessible.			
Discontinue any nonessential construction of installation activities.			
Ensure that trash containers are emptied, and aisles and walkways are clear.			
Ensure pest controls are up to date. There should be no signs (present or past) of pests or rodents.			
Ensure all chemicals are identified and labeled with an MSDS (material safety data sheet).			
Ensure all materials are labeled and within expiration dates.			
Ensure materials are properly identified. Ensure that nonconforming product is properly identified, segregated, and dispositioned per requirements.			

(Continued)

Continued

When an Inspection Is Announced	Yes	No	Follow-Up Required
Ensure special storage requirements are in place, if applicable.			
Ensure no uncontrolled documents, sticky notes, crib sheets, unauthorized changes are in use.			
Ensure records, routers, travelers, etc. are in place and correctly filled out.			
Ensure operators are only performing operations for which they are trained and have documentation.			
Ensure no obsolete or superseded documents are in use.			
Ensure all procedures and documents are available at point of use. Review procedures and documents for GDP errors.			
Ensure nonfunctioning equipment is tagged.			
Ensure any equipment under validation is identified as not-for-production use.			
Ensure all calibration stickers are in place, legible, and within calibration.			
Ensure preventive maintenance is up to date.			
Ensure environmental recorders (Example: Temperature) are in place, in specification, and accurate.			
Ensure software validation documentation is available for all software-controlled equipment, databases, manufacturing execution systems, etc.			
Ensure all nonproduction areas and cleanrooms are identified and segregated.			
Plant Tour—Clean Rooms			
Ensure cleaning logs are up to date.			
Ensure the gowning area is clean, orderly, and signs are in place.			
Review and reinforce procedures for gowning, face masks, hair covers, hand washing, jewelry, make-up, etc. Make sure that investigators comply with standard procedures.			
Ensure gowning and washing steps are clearly posted.			
Ensure gowning procedures are consistently followed. Note: Investigators may take time to watch people and make sure they are following procedures.			
Ensure pass-throughs and other transfer areas are clean and functional.			
Ensure cleanroom gages are functional, calibrated, and reading in specification.			
Ensure door interlocks are functional.			

(Continued)

Continued

When an Inspection Is Announced	Yes	No	Follow-Up Required
Ensure continuous monitoring equipment is functional.			
Ensure pest controls are up to date.			
Ensure there are no obstructions of airflow.			
Ensure any service providers (including cleaning personnel) used are on the approved supplier list.			
Plant Tour—Calibration Area			
Ensure area is neat, clean, and orderly.			
Ensure standards are properly stored and in specification.			
Ensure certifications are on file for any outside calibrations.			
Ensure environmental controls are identified and in specification.			
Ensure outside calibration service providers are on the approved supplier list.			
Ensure all nonfunctional equipment is identified and segregated.			
Ensure area is clean and orderly and packaging materials are removed promptly.			
Ensure all material handling racks are neat, clean, and orderly.			
Ensure all products and materials are all properly labeled.			
Inspection equipment is calibrated.			
Ensure all rejected materials are identified, segregated, and controlled.			
Ensure MRB (material review board) area (or cage) is properly controlled with restricted access.			
Ensure nonconforming material is promptly identified, segregated, and dispositioned.			
Ensure all environmental controls are in place and monitored.			
Plant Tour—Stores/Distribution			
Ensure pest controls are in place and current. Ensure there are no signs of rodents or pests, past or present.			
Ensure accepted materials are properly labeled.			
Ensure environmental controls are monitored and acceptable.			
Plant Tour—Sterilization			
Ensure area is neat, clean, and organized.			
Ensure sterile and nonsterile products are segregated and clearly identified.			

(Continued)

Continued

When an Inspection Is Announced	Yes	No	Follow-Up Required
Ensure rejects are properly and promptly segregated, identified, and controlled.			
Ensure equipment maintenance and calibration is current and acceptable.			
Ensure operator training and certifications are up to date.			
Ensure sterilization validation documentation is available.			
Ensure process revalidation activities are documented.			
Ensure routine bioburden and endotoxin test programs are in place. Reports demonstrating acceptable levels are available.			

Design Controls

	Yes	No	Follow-Up Required
Prepare to present procedures and, if necessary, an overview of design control and risk management.			
Be prepared for the FDA to select a design project that challenges the design control process. This could be a product that was involved in a recall, MDR, has high risk or new technology. Or it may simply be your latest product launch.			
Show design and development plans. Be prepared to show that the design plan was updated as the project evolved.			
Show that risk analysis (i.e., risk management) is included in the plan.			
Show evidence of design inputs and requirements.			
Show evidence of design outputs and specifications.			
Show evidence that design outputs essential for proper functioning of the device are identified.			
Show design verification protocols, data, reports. Show that design verification confirmed that outputs met input requirements.			
Show that acceptance criteria were established prior to performance of verification and validation.			
Show design validation protocols, data, reports. Confirm that design met user needs and intended uses. Demonstrate that validation included testing under simulated or actual use conditions.			
Has software been validated?			
Show traceability that design outputs met design inputs.			
Has design transfer been completed?			
Show design review meeting minutes and action items. Show evidence action items have been completed.			

(Continued)

Continued

When an Inspection Is Announced	Yes	No	Follow-Up Required
Show that design validation did not leave any unresolved discrepancies.			
Show software validation, if applicable.			
Show that risk analysis was performed. Note: The QSR (Quality System Regulation) uses the term "risk analysis" but risk management per ISO 14971 is the recognized international standard.			
Demonstrate that design validation was accomplished using initial production devices or their equivalents.			
Be prepared to show that design changes were controlled including verification and validation.			
Show evidence that design reviews were conducted. Ensure design reviews account for risk analysis and change control. Be prepared to show meeting minutes and an independent observer.			
Show that the design was properly transferred to production. Be prepared to show the DMR.			
Production and Process Controls			
The FDA will select a process for review that it perceives as high risk based on CAPA indicators of problems, use of the process for high-risk devices, degree of risk for causing failures, lack of familiarity, use of the process for multiple products, or other criterion for the assignment.			
Be prepared to provide procedures for the process selected. These may include procedures for in-process and finished device acceptance activities.			
Be prepared to provide procedures on environmental and contamination controls as appropriate.			
Be prepared to show examples of verification activities necessary to support validated processes, such as review of process parameters, dimensional inspections, performance tests, etc.			
Be prepared for the FDA to determine if control and monitoring activities are in compliance with the DMR. They may review shop floor documents, batch records, or observe activities.			
Be prepared for the investigator to note the identification number of a significant piece of manufacturing equipment and a piece of inspection equipment. Be prepared to provide records for maintenance, calibration, and adjustment of the equipment.			
If using external suppliers for maintenance, calibration, adjustment, cleaning, etc. be prepared to show how			

(Continued)

Continued

When an Inspection Is Announced	Yes	No	Follow-Up Required
the supplier is evaluated, monitored, and included on the approved supplier list.			
Be prepared to provide a list of batch records.			
Be prepared to show receiving acceptance and purchasing data for at least one component of raw material.			
Be prepared to provide records of environmental and contamination controls.			
Be prepared to discuss any nonconformances.			
Be prepared to show validation [IQ (Installation Qualification), OQ (Operational Qualification), and PQ (Performance Qualification)] protocols, data, and reports.			
Be prepared to show that instruments used in validation were properly calibrated and maintained.			
Be prepared to show that predetermined specifications were established.			
Be prepared to show that sampling plans were based on statistically valid rationale.			
Be prepared to show evidence that predetermined criteria for success were consistently met.			
Be prepared to show that process tolerance limits were appropriately challenged.			
Be prepared to show that process operators are appropriately trained and qualified.			
Be prepared to show that software to control the process is validated.			
Purchasing Controls			
Be prepared to show procedures for purchasing control, supplier evaluation, and receiving inspection.			
Be prepared to show requirements for quality agreements.			
Be prepared to show that you require suppliers to notify you of any changes to purchased materials.			
Be prepared to show methods for supplier monitoring including remaining on or removing from the ASL (approved supplier list).			
Be prepared to show the ASL, including critical suppliers.			
Be prepared to show a listing of SCARs (supplier corrective action requests) for the last 2 years (or the time since the last FDA inspection).			
Be prepared to show validation and/or receiving inspection of critical materials.			

(Continued)

Continued

When an Inspection Is Announced	Yes	No	Follow-Up Required
Be prepared to show purchasing data and records.			
Be prepared to show the supplier audit schedule.			
Be prepared to show supplier risk ratings.			
Be prepared to show a list of CAPAs related to supplier issues.			
For-Cause Inspection—Recall or Field Corrective Action			
Be prepared to show Physician Notification(s) and evidence of due diligence and completeness of contact.			
Be prepared to show the Risk Assessment (HHE or Health Hazard Evaluation).			
Be prepared to show the associated CAPA file.			
Ensure reconciliation of all recalled product is complete and accurate. Evidence of destruction of recalled product must be available.			
Ensure device tracking data (if applicable) has been updated completely.			
Be prepared to show the complaint history and associated MDRs that lead to a corrections and removals decision.			
Pre-PMA Inspection (Product Focused)			
Be prepared to show the design history of the applicable product.			
Be prepared to show the process qualification and validation of the applicable product.			
Be prepared to show clinical studies or other product performance data.			
Post-PMA Inspection (Product Focused)			
Be prepared to review design changes and validation since product release.			
Be prepared to review process changes and validation since release.			
Be prepared to review analysis of returned products.			
Be prepared to review complaints and MDRs for the applicable product.			
Be prepared to review any CAPAs for the applicable product.			

(Continued)

Continued

When an Inspection Is Announced	Yes	No	Follow-Up Required
During the Inspection			
Conduct daily debriefs with management per your communication plan.			
Conduct daily debrief with inspection support team to prepare for the next day.			
Send daily notes per your communication plan.			
Keep identical duplicates of all documents provided to the FDA.			
Require investigators to comply with all applicable policies and procedures while on company premises. Pay attention to gowning and clean room procedures.			
The investigator should be escorted at all times.			
If samples are taken by the FDA, request duplicates.			
Restrict photography and recording per your established policy. Consult higher level management or legal counsel before making any exceptions.			
Train and prohibit employees from signing, listening to, or acknowledging any affidavits from the FDA. Consult legal counsel for guidance if necessary.			
Preparing for a Close-out Meeting with the FDA (if you anticipate a Form 483)			
Identify appropriate personnel for closing meeting. Include the highest level of management at the site, highest level site quality and compliance, inspection leader, and the individual who will be responsible for leading the Form 483 response.			
Consider options for annotating Form 483 observations: Reported corrected, not verified; corrected and verified; promised to correct; and under consideration. If there is any concern about the legitimacy or accuracy of an observation, consider annotating with "under consideration" or not annotating at all.			
If an observation is unclear, ask the investigator for clarification.			
Remember to ask the investigator if all questions have been answered to their satisfaction. Ask if there are any remaining concerns for management discussion.			
Ask the investigator how they will classify the inspection: NAI (no action indicated), VAI (voluntary action indicated), or OAI (official action indicated).			

(Continued)

Continued

When an Inspection Is Announced	Yes	No	Follow-Up Required
Preparing a Form 483 Response			
Identify an owner responsible for overseeing the response process and making decisions about overall content and strategy. Small companies without experience in this should include experienced outside consultation.			
Develop a response timeline to allow for a complete, accurate, and timely response. Allow time for review and comment.			
Activate a response team. Assign an owner for each Form 483 observation.			
Complete and approve thorough responses (see below).			
Allow adequate time for review and administrative tasks.			
Use the CAPA process to manage Form 483 responses.			

You may or may not receive advanced notice of an FDA inspection. Sometimes the FDA investigator just shows up at the door. It is important that you prepare the receptionist or security personnel for this. Sometimes, the investigator may call up to 5 days in advance to notify you of an upcoming inspection. Foreign manufacturers receive up to 2 or 3 months' notice. You may be requested to send information such as a quality policy and quality system manual or equivalent for preinspection review.

When the investigator arrives, they will ask to see top management. They should present their credentials and issue a Form 482, Notice of Inspection, which explains their legal authority to inspect.

During the inspection, the investigator may ask to tour the facility. This is a common practice and provides the investigator with additional information to plan and conduct the inspection. They may select one or two random people and ask them what the quality policy is. They may jot down some equipment numbers and/or gage calibration numbers depending on the scope of their inspection.

Personnel that have to interact with the investigator or go into the front room are often nervous. This is understandable. Most FDA investigators are skilled at questioning and have developed strong interviewing techniques to get more information. Some investigators will stop talking and just look at the interviewee. This causes the interviewee to become more nervous and babble on to fill the void. Do not babble. It is acceptable for personnel to pause. Other investigators will deliberately needle and pressure interviewees to cause frustration and a resulting spew of information. Do not respond. Share the following "Dos and Don'ts" as part of your training for SMEs.

Do	Don't
Do answer truthfully and give complete responses.	Do not ever lie or mislead.
Do stick to facts and refer to SOPs, procedures, etc., as needed.	Do not guess or discuss opinions.
Do inform the inspection leader if a question is outside your area of responsibility. The inspection leader can bring in additional SMEs as appropriate.	Do not respond outside your area of expertise.
Do express legitimate concerns.	Do not disagree or engage in nonconstructive arguments.
Do answer accurately and completely.	Do not provide unsolicited information or documents.
Do ask the investigator to be specific if their question is unclear.	Do not answer "what if" or other hypothetical questions. Ask for clarification.
Do pause as needed.	Do not feel compelled to fill the void by talking.

Some organizations coach their SMEs to answer only with a yes or no. They coach SMEs to provide as little information as possible and to force investigator (s) to dig for every little bit of information. This is bad advice as it creates a confrontational attitude with the investigator. It is better to show cooperation. That does not mean volunteering unnecessary information or babbling on. But, do give complete responses. If the investigator asks if you have a procedure for XYZ, do not just say "yes" and force them to ask for the procedure name or number. It is okay to say, "yes, procedure 123."

If personnel accidently give incorrect information, it should be discussed with the backroom leader or inspection leader for correction and communication with investigator(s). Do not allow incorrect information to go unaddressed.

Attend to the needs of your personnel. An inspection can be very stressful and require long, hard hours of work. Provide water, coffee, and soda. Bring in refreshments. Have snacks or meals ready for personnel that are coming in early and staying late to manage your inspection. Cheese puffs and plenty of chocolate seem to be much appreciated favorites during an FDA inspection!

You may also offer water and coffee to investigators. Keep some refreshments in the front room. Investigators are not your enemy. They just have a job to do. Treat them professionally. Investigators are not allowed to accept anything of value but do appreciate water, coffee, and may even munch on a bagel or a cookie. Others will be stricter about rules and will leave your facility to go out to lunch (or take a break) on their own. Do not attempt to offer them anything of value.

Good communication is vital during an inspection. The inspection leader should have a daily debrief meeting with the inspection team to communicate status, identify potential issues, needed information or actions before the next day. Identify support needed from other sites, SMEs, or other resources. Take time to coach and prepare SMEs for the next day.

Prepare for a closing meeting with the FDA on the last day of the inspection. Appropriate attendees include the highest level of site management and the highest level of site quality management (or management representative if not the inspection leader). If you suspect a Form 483 will be issued, make sure the individual responsible for preparing the response is included. They will need to hear the conversation first hand in order to draft a complete and thorough response.

The FDA will use Turbo EIR, their method for linking observations to specific regulatory requirements. Turbo EIR provides more uniform Form 483s and EIRs and helps the FDA analyze information. Note: The FDA can only enforce what is specifically required by regulation. For example: The regulations use the term risk analysis (21 CFR 820.30 (g)) even though they consider ISO 14971 Risk Management as an accepted international standard (this is explained in the preamble). Because of this, citations will be made against risk analysis to the effect of "risk analysis was not performed" or "risk analysis was inadequate."

Sometimes the FDA investigator will attempt to issue an affidavit to document evidence collected at the manufacturer. An affidavit is of no value to you and may be used against you at a later date. Most companies train their personnel not to listen to, read, acknowledge, or sign an affidavit. Consult your legal counsel ahead of time so that you have an established policy and are prepared for this.

A Form 483 (Notice of Observations) may be issued at the conclusion of the inspection when the investigator(s) have observed any conditions that, in their opinion, constitute violation(s) of the Food Drug and Cosmetics Act. Observations will be made when there are indications that the product is adulterated or misbranded. The Form 483 notifies management of the objectionable conditions. The Form 483 is discussed during the closing meeting. Each observation is read and discussed.

Management can, and should, ask for clarification if necessary. The investigator should ask if you want to annotate the Form 483. It is possible to annotate observations as:

- Reported corrected, not verified
- Corrected and verified
- Promised to correct
- Under consideration.

If at all possible, correct items while the investigator is there, and you can provide proof of correction. This can minimize the severity of future enforcement actions from the FDA. A willingness to promptly correct issues is always positive. If you are unsure of an observation or concerned about the accuracy, you may choose not to annotate it or to annotate it as under consideration.

Management should ask the investigator(s) if all their questions have been answered satisfactorily and if there are any additional concerns not included in the Form 483. It is important to understand if there are any other issues that might later be raised in the EIR that could be used as the basis for further enforcement action (i.e., warning letter).

At the end of the inspection, ask the investigator what classification they will recommend for the inspection. Classifications are

- NAI—No action indicated.
- VAI—Voluntary action indicated. This means some deficiencies identified but nothing very serious.
- OAI—Official action indicated. This means serious issues were identified and the FDA must take action to assure correction.

> **NOTE**
>
> Any violations of medical device reporting (MDR), device tracking, and corrections and removals regulations are considered very significant, will prompt warning letter consideration, and will require concurrence from CDRH (Center for Devices and Radiological Health).

After the inspection, conduct a postinspection debriefing with the entire inspection support team. Identify positives and opportunities for improvement. Openness and transparency should be encouraged. Archive inspection documentation for future reference. Refill the supplies box, seal, and store it for the next inspection. Create a reminder to follow-up and make sure you get an EIR in about 60 days after the inspection.

The inspection leaders should also document their own private thoughts on how the inspection went. What was done well and what could have been done better? How did internal personnel perform? Jot down your ideas while they are fresh in your mind. They will become the basis for improvements to the next inspection preparedness plan. Be sure to update the plan and prepare for the next inspection.

If a Form 483 is issued, a response is required within 15 days of the inspection (generally from the day the Form 483 was issued). A complete, accurate, and thorough response is incredibly important to prevent further enforcement action. It is your opportunity to convince the FDA of the adequacy of your actions. The FDA will take your response into account when determining whether to take further action. Factors that can affect issuance of a warning letter include:

- A firm's history
- The nature of the violation
- The risk associated with the product
- The overall adequacy of the firm's response
- Whether documentation of corrective actions was provided in the response
- Whether the actions and timeframe are commensurate with the risk
- Progress on corrective actions is reasonable.

Make sure you plan a response timeline so that you can adequately determine root cause, identify actions, and draft the response. Allow adequate time to review the response (with multiple reviewers and rounds), and complete administrative tasks such as copying and mailing.

Although the inspection team may be feeling exhausted after the inspection, there is still important work to be done. Do not underestimate the importance of a good response! Every warning letter from the FDA contains the words "We have reviewed your firm's response and conclude that it is not adequate because...." Some frequently cited reasons are:

- Your firm's response did not include supporting documentation for all its planned corrections and corrective actions.
- Your response did not address this deficiency.
- Your response did not evaluate the potential impact of these violations on distributed devices and take steps to mitigate the risks as needed.
- You have not retrospectively evaluated XXX.
- You have failed to consider systemic corrective actions.
- You have failed to provide evidence of corrections, corrective actions, and systemic corrective actions.
- While an updated procedure was submitted in the response, there is no evidence of training.
- You have failed to provide evidence of implementation.
- In addition, you did not provide evidence that you documented the changes.
- There is no evidence you retrospectively reviewed previous items to ensure they are completed, as required.
- Your response did not include XXXX (a specific part of the regulation).
- Your response did not address this observation.
- The adequacy cannot be determined at this time.
  - Please provide the results of completed CAPA XXX, including supporting evidence, as the data become available.

The lessons learned here are that a response needs to be created for each and every observation on the Form 483. It must include identifying root cause and making sure procedures (see Chapter 3, Establish and Maintain, for more on establishing and maintaining, and Chapter 4, QMS Structure, for writing good SOPs) are brought to full compliance. You must include evidence that personnel were trained and procedures were implemented. You must show evidence of a retrospective review to identify and correct further instances of the same non-conformity. You must assess the impact of any violations on product in the field and mitigate risk if appropriate. Sometimes, adequacy cannot be determined right away. In that case, the FDA will plan for a follow-up inspection at some point in the future. Depending on the number and severity of observations, a complete and thorough response can prevent further action (i.e., warning letter) from the FDA. Work with an experienced consultant to help draft a good Form 483 response.

Elements of creating a robust Form 483 response include the following:

- Identify the root cause(s) to determine appropriate actions to address the issue.
- Determine comprehensive actions that:
  - Correct the specific examples identified in the observation
  - Identify and correct additional or similar examples to those specifically identified
  - Prevent or mitigate the risk of future occurrence
  - Address systemic issues.
- Provide clear timelines of actions.
- Evaluate if the observation has impact on product in the field or distribution chain. Consider further necessary actions such as product hold orders and field corrective actions. If the assessment has not been completed, the response should indicate when it will be completed.
- Consider addressing any further concerns discussed during the close-out meeting.
- For foreign inspections pay very close attention to any issues involving 21 CFR 803, 21 CFR 806, 510(k) clearance, and PMA approval.
- When changing SOPs or other procedures, highlight the changed sections to show observations have been addressed.
- If appropriate, provide an overall plan of comprehensive actions and dates.
- Include training records to show procedures have been implemented.
- Include all supporting documents. Prepare a list of attachments with reference numbers, version numbers, dates, etc., to provide clarity as needed.

Your response should also include a cover letter that:

- Expresses the commitment of the company to comply with all applicable laws and regulations.
- Addresses any concerns about management responsibility.
- Commits to providing an update at a designated time in the future. The date should show needed urgency but must be realistic.
- Instills confidence that you have a thorough, detailed plan.
- Indicates that all documents are marked as company confidential.
- Includes contact information.

Use the rigor of your CAPA system to manage observations, and document responses and commitments. Execute actions precisely and review status during management review. Management with executive responsibility should be actively aware of responses to regulatory inspections. After all, they are responsible for a suitable and effective QMS with adequate resources.

> **TIP**
>
> Do not consider the Form 483 as the entire history of the inspection. Be sure to follow up and review the EIR. The EIR is completed by the investigators after they return to the office. It contains the details of the inspection including a narrative explaining who was interviewed, what documents were reviewed, and linkage to the Form 483.

The EIR will first go to a supervisory investigator at the FDA who makes an initial determination of the seriousness of the observations and recommends a determination of NAI (no official action indicated), VAI (voluntary action indicated), or OAI (official action indicated). From there the EIR will go to the compliance officer (a separate part of the district office) who may choose to accept or not accept the recommendation. Factors that impact the decision include:

- Significance of the observations
- History of the site and other company sites
- Degree to which the evidence supports JIVR
- Compliance by the FDA investigator with the legal obligations and limitations the FDA operates with during inspections.

> **HINT**
>
> If a company can show the inspection was not in compliance with FDA obligations, then it may be possible to avoid a warning letter but only do this with careful consideration and discussion with experienced legal counsel.

The compliance officer will review the company response to the Form 483 to determine if it is timely, effective, addresses systemic issues, and ensures prevention of recurrence. If the compliance officer determines a warning letter or further action is warranted, the EIR may undergo further review by the district office director or other SMEs. Depending on the type of violation, the EIR may be reviewed by the Office of Compliance at CDRH. If judicial action (seizure, consent decree, injunction) is considered then other parties such as FDA Office of Chief Counsel, and U.S. Attorneys or Department of Justice officials may become involved.

The FDA has a "Recidivist Policy" for companies that have a pattern of correcting violative conditions in response to a warning letter only long enough to pass a follow-up inspection. If the FDA observes the violations again in a subsequent inspection, it may trigger a recidivist warning letter. The recidivist warning letter contains additional language requiring annual certification by an approved third party. The FDA can use additional enforcement methods, if necessary, such as seizure, injunctions, civil money penalties, and prosecution.

Most companies consider FDA inspections much more rigorous that other types of regulatory inspections. If you have an inspection preparedness process as described above, you should be able to easily manage other inspections. The difference is the regulatory requirements that you are being inspected against. Make sure you clearly understand the requirements that you are being inspected against.

In conclusion, an FDA inspection is a very significant event. Because it is so important, it can be very stressful. But, the stress can be eased by being well prepared. Inspection preparedness requires planning, preparation, and practice. Start your planning with awareness of past inspection results and commitments. Take time to create a written plan covering the items discussed above. Practice the elements of your plan. You will be rewarded when the FDA arrives, and you can properly present your suitable and effective QMS!

Conclusion

The goal of this book is to provide a road map to guide you in creating an effective and efficient quality management system (QMS). It is to help you develop a QMS that is nimble and flexible, but still meets regulatory requirements and ensures quality outcomes for your customers. This requires:

- Understanding the regulations and how to translate and establish them into the structure and processes of your QMS
- Clear QMS structure, processes, procedures, and supporting infrastructure
- Clear roles and responsibilities for management with executive responsibility, the management representative, process owners, and all individuals in the company
- Alignment of all the processes, metrics, personnel, and infrastructure
- Essential capabilities or MEDICS (Monitor, Embrace, Define, Identify, CAPA, Share)
- Tools for process monitoring and improvement.

Throughout this book, we have covered the regulations, interpretation, necessary capabilities, and methods needed to successfully implement them. We have identified useful tools necessary for defining, implementing, and training the organization. Tools for process monitoring, statistical techniques, root cause analysis, failure investigation, and corrective and preventive action have been identified. Note: It is impossible to comprehensively cover all these tools in one (or even many) books. I recommend you review these tools in greater depth!

It is a never-ending quest to break a cycle of ineffectiveness and inefficiency. An ineffective process will never yield predictable results. There will always be unpleasant surprises causing delays, rework, and potentially serious compliance and quality problems. In this book, I explored methods for reducing these problems and shifting from a mode of reaction to prevention. Ultimately, this is one of the key benefits that the quality and compliance organization brings to the medical device company.

If you are an experienced quality and compliance professional, you have undoubtedly previously encountered many of the concepts in this book. Hopefully you have gotten a refresher and found some additional detail or nuance in key concepts. You may have found some new areas of improvements and techniques for evaluating MEDICS capabilities. You should be able to define your current business situation and level of maturity. The goal for you is to connect the dots and be a change agent for driving improved quality outcomes.

Quality leaders have some new tools to earn a seat at the table that will enable your company to use quality and compliance as a competitive strength. You have concepts to change thinking from the cost of quality to the value proposition of quality. Incorporate that into your vision and strategy for improvement. No matter what your current business and quality situation is, you can create a vision and a strategy to reach that future state.

If you are management with executive responsibility, you should have a new-found respect and awareness of the need for regulation. You have learned about how specific issues and problems shaped the regulations industry must work within today. You now know the responsibilities of your role to ensure a suitable and effective QMS. You are equipped with tools to interpret management review information to understand quality and compliance risk. This equips you to make risk-based decisions and ensure adequate resources. Creating a culture of quality, data-driven information, and customer focus is not a cost but rather an investment in prevention. And that is good for your business.

If you are a functional or process owner, you should have more awareness of how to translate regulatory requirements to your business process in a seamless manner. You are equipped with better understanding of the process approach. You are now prepared to fully define, create, and monitor process performance. You have tools for evaluating process maturity and reaching process excellence. Quality and compliance are not a burden but an investment in making your process more consistent and predictable. And predictable processes result in better business results.

Armed with this set of information, as an organization you can establish, maintain, and improve your QMS! Best wishes.

References

American Medical Association, 1911. Nostrums and Quackery: Articles on the Nostrum Evil and Quackery. Press of the American Medical Association, Chicago.

Breyfogle III, F.W., 1999. Implementing Six Sigma. John Wiley and Sons, New York.

Crosby, P.B., 1980. Quality Is Free, the Art of Making Quality Certain. Penguin Publishing Group, New York.

Crosby, P.B., 1984. Quality Without Tears. McGraw-Hill Book Company, New York.

FDA, 1997a. Medical Devices; Current Good Manufacturing Practice (CGMP) Final Rule; Quality System Regulation. Federal Register October 7, 1996. Volume 61.

FDA, 1997b. Medical Device Quality Systems Manual: A Small Entity Compliance Guide. Food and Drug Administration, Rockville, MD.

FDA, 1999. Guide to Inspections of Quality Systems. Food and Drug Administration, Center for Devices and Radiological Health, Rockville, MD.

FDA, 2011. Understanding Barriers to Medical Device Quality. Food and Drug Administration, Center for Devices and Radiological Health, Rockville, MD.

FDA, 2015. Compliance Program Guidance Manual 7382.845 Inspection of Medical Device Manufacturers. Food and Drug Administration, Rockville, MD.

FDA, 2017a. Investigations Operations Manual. Food and Drug Administration, Rockville, MD.

FDA, 2017b. Regulatory Procedures Manual. Food and Drug Administration, Rockville, MD.

Global Harmonization Task Force, SG3, 2010. Quality Management System – Medical Devices – Guidance on Corrective and Preventive Action and Related QMS Processes.

Global Harmonization Task Force, SG3, 2012. Quality Management System – Medical Devices – Nonconformity Grading System for Regulatory Purposes and Information Exchange.

Global Harmonization Task Force, SG4, 2007. Guidelines for Regulatory Auditing of Quality Management Systems of Medical Device Manufacturers – Part 3: Regulatory Audit Reports.

Global Harmonization Task Force, SG4, 2008. Guidelines for Regulatory Auditing of Quality Management Systems of Medical Device Manufacturers – Part 1: General Requirements.

Global Harmonization Task Force, SG4, 2010a. Guidelines for Regulatory Auditing of Quality Management Systems of Medical Device Manufacturers – Part 2: Regulatory Auditing Strategy.

Global Harmonization Task Force, SG4, 2010b. Guidelines for Regulatory Auditing of Quality Management Systems of Medical Device Manufacturers – Part 4: Multiple Site Auditing.

Global Harmonization Task Force, SG4, 2012. Guidelines for Regulatory Auditing of Quality Management Systems of Medical Device Manufacturers – Part 5: Audits of Manufacturer Control of Suppliers.

Haggar, B. (Ed.), 2013. The Biomedical Quality Auditor Handbook. American Society for Quality, Quality Press, Milwaukee.

International Standard, ISO13485, 2016. Quality Systems — Medical Devices — Requirements for Regulatory Purposes. International Organization for Standardization, Geneva, Switzerland.

International Standard, ISO14971, 2007. Medical Devices — Application of Risk Management to Medical Devices. International Organization for Standardization, Geneva, Switzerland.

Jones, P.A. (Ed.), 2013. Fundamentals of US Regulatory Affairs. eighth ed. Regulatory Affairs Professionals Society, Rockville, MD.

Juran, J.M., 1999. Juran's Quality Handbook, fifth ed. McGraw-Hill, New York, NY.

Kausek, J., 2006. The Management System Auditor's Handbook. American Society for Quality, Quality Press, Milwaukee, WI.

Pyzdek, T., 2001. The Six Sigma Handbook, A Complete Guide for Greenbelts, Blackbelts, and Managers at all Levels. McGraw Hill, New York, NY.

WEBSITES

Advanced Medical Technology Association (Advamed). www.advamed.org/

American Society for Quality (ASQ). http//www.asq.org

Association for the Advancement of Medical Instrumentation (AAMI). www.aami.org/

Food and Drug Administration. http//www.fda.gov

International Medical Device Regulators Forum. http//www.imdrf.org

International Organization for Standardization. http//www.iso.org

Medical Alley Association. https://www.medicalalley.org/

Medical Device and Diagnostic Industry. https://www.mddionline.com/

Medical Device Innovation Consortium. http//www.mdic.org

Medical Device Manufacturers Association. http://www.medicaldevices.org/

Plain Language Action and Information Network (PLAIN). www.plainlanguage.gov/

Index

Made in the USA
Coppell, TX
16 January 2020